J. Spencer

Middleburg

1978

4

Reaching the Peasant Farmer

David K.
Leonard

Reaching
the
Peasant
Farmer

Organization Theory and
Practice in Kenya

The University of Chicago Press
Chicago and London

The University of Chicago Press
Chicago 60637
The University of Chicago Press, Ltd.
London

DAVID K. LEONARD is assistant pro-
fessor of political science at the University of California at Berkeley. He is
the editor of *Rural Administration in
Kenya: A Critical Appraisal*.

Library of Congress Cataloging in Publication Data

Leonard, David K
    Reaching the peasant farmer: organization theory and
practice in Kenya.

    Includes bibliographical references and index.
    1. Agricultural extension work—Kenya.   2. Peasantry
—Kenya.   I. Title
S544.5.K4L45   630'.7'15   77–1823
ISBN 0–226–47260–4

For Rowland
and
Mary Jane Leonard

Contents

Part 3
Organizational Structures
and the Performance of
the Individual Agent

Illustrations

Tables

Preface

African problems can be summed up in one word: underdevelopment.... Without industry, without resources of energy except manpower, without modern methods of cultivation and consequently high yields, without buying power for the products of future industries, a new vicious circle forms. It should be attacked by industrialization, certainly, but it can be broken more effectively and quickly through agriculture. Progress in agriculture should not be considered as a preliminary to industrialization, but as an indispensable corollary.[1]

These words of René Dumont have proved prophetic, not only for Africa but for the entire developing world. Rural development has become the focus of attention for the 1970s. At the very center of the ensuing concern and activity is the small farmer and his productivity. A few countries have chosen to deal with him through the collectivization of agriculture and another small number implicitly have decided to ignore (and thus destroy) him while concentrating on large-scale capitalist farming. But the overwhelming majority of developing countries are committed to transforming the economic conditions of farmers who are and will remain the managers of their small, individual plots of land. Targets—such as those of the World Bank—to increase small farmers' net incomes by 5 percent per annum are now commonplace; yet these targets are unrealistically optimistic given present levels of knowledge and organizational capacity in the developing world.

Small-scale agricultural development has a large number of prerequisites. Agricultural technologies must be invented that will be suited to the amounts of land, labor, and capital available to small producers. Crop prices must be high enough to provide real incentives to produce. Marketing structures and roads must be created to get harvests to the selling points and fertilizers and seeds to the farmers. Many countries still have not dealt with these essentials, and the great majority need to make further improvements. Yet experience shows that even when all these prerequisites are met, rapid small-farm development does not always take place. News of new production possibilities has to be disseminated and instruction in new animal and crop technologies provided to the small farmers.

Reaching the small farmer and helping him to develop has been and will remain the responsibility of government extension services. This book will demonstrate that these responsibilities are not being met at present. Our purpose, however, is not to point to failure. Rather, we mean to analyze the organizational problems of government extension organizations and to demonstrate how they can be solved. By so doing we hope to show how the last link in the chain reaching the small farmer can be forged and rural development promoted.

Admittedly, the organizational problems of extension services is one part only of the process of creating small-farm development; but it is the one part on which we lack a systematic literature, and it is the aspect that constitutes the most important remaining constraint on rural development in a number of countries. As Uma Lele remarks in a review of rural development programs in Africa, "All too often the administrative systems used in the delivery of services and the manpower they require seem to be a far greater constraint . . . than are finances."[2]

Superficially, this book is a case study of the organizational problems of agricultural extension in Kenya—one African country with its own unique characteristics. At a deeper level, it is actually an analysis and prescription for extension organizations of all types throughout the developing world. We dare to claim such a wide application for our work for two reasons: first, Kenya is an ideal case for isolating and studying these problems; and second, we have placed the study within a framework of organization theory which has been widely tested and therefore permits broad generalizations.

Kenya is a strategic country for analyzing the organizational problems in extension because it has already met the other prerequisites of agricultural development for its most important agro-economic zones. It has a good agricultural research establishment and a number of tested and profitable crop innovations to offer to the country's small farmers. Kenya's market prices provide a strong incentive to the small farmer (although they do still mask elements of exploitation).[3] The supply and marketing infrastructure in many of the small-farm areas of the country is exceptionally good for a developing state. Finally, Kenya is fortunate in having a ratio of extension workers to farmers of 1:310, one of the highest in the developing world.[4]

Despite all these advantages the Kenyan extension services have been disturbingly inefficient. For example, a new and highly profitable hybrid corn technology was made available to Kenyan small farmers in 1963. Eight years later, only 19 percent of the small-farm corn acreage was sowed with hybrid seed.[5] Subsequent experiments in Kenya have demonstrated that properly organized extension efforts can produce adoption rates near 100 percent.[6] The experiments of Ascroft and his associates and of Schönherr and Mbugua involved closer coordination in the provision of seed, fertilizer, and information to the farmer, careful planning for the flow of technical information through the extension service hierarchy, involvement of junior extension agents in planning their own work, and use of group extension methods to facilitate both the diffusion of information and the supervision of junior staff. This book deals with precisely these practical organizational issues and others like them, drawing on the exceptionally large amount of investigative and experimental research on Kenyan agricultural extension and administration in recent years.

We feel able to generalize from our findings because of our rigorous use of organization theory. This body of theory contains a large number of practical, tested propositions about worker motivation, ways to stimulate feedback and innovation, effective supervisory styles, and the like. When we began our research we did not know whether these propositions, based largely on studies in North America and Europe, would be applicable in Kenya. In the course of our work, however, we found that particularly those propositions about organizational behavior derived from exchange theory are valid in Kenya. This strongly suggests that they are valid elsewhere in the developing world as well. Therefore we have placed our research findings in an exchange and organization theory framework. In the process of analyzing the detailed findings of our research, we introduce the basic theoretical perspectives and specific propositions of organization theory that we believe are useful to the extension practitioner. We hope that those not now acquainted with organization theory will learn through this book how helpful it can be to their work. The specific facts related here are unique to Kenya and cannot be generalized, but the theories that have enabled us to analyze and correctly predict the structure of these facts can be applied generally. These theories and the ways in which we have used them will assist in diagnosis and prescription for unique extension organizations in other developing countries.

Of course, the application of organization theory that gives our findings wide usefulness to practitioners also makes our work of value to students of organizational behavior. For we do not apply organization theory uncritically. We have found it necessary to choose among its competing streams, to modify it, and to extend it. In so doing we hope we have made a contribution to the task of making organization theory genuinely cross-cultural. We also trust that we have added to our systematic understanding of one more important aspect of African administrative behavior.

This study is the product of many different minds. The actual writing is mine, of course, but that was only a small part of the process of its creation. First of all, I must acknowledge the invaluable contribution of my research assistants—Humphries W'Opindi, Jack K. Tumwa, Edwin A. Luchemo, and Bernard Chahilu, all students at the University of Nairobi when they worked for me. They conducted most of the interviews on which this study is based and processed the data. More important, however, they made this research project a part of their own intellectual interests and shared with me innumerable insights concerning the workings of rural Kenya. The use of the pronoun *we* rather than *I* throughout this volume acknowledges the difficulty I have in separating their findings and thoughts from my own.

Less directly, this study is the product of the intellectual communities of the universities of Chicago and Nairobi. At Chicago, Peter Blau (now of Columbia), Kenneth Prewitt, Lloyd Rudolph, and Aristide Zolberg provided extremely valuable comments on the first drafts of this volume, although I have sometimes been unsuccessful in following their advice.

The University of Nairobi contributed at several levels. Its Institute for Development Studies employed me from 1969 to 1971 while I was doing my fieldwork and paid for most of the costs of data collection and analysis. (The Center for the Comparative Study of Political Development of the University of Chicago also provided funds for some of the computer work.) Equally important, the institute provided the intellectual framework within which my specific research project was formed, and its program of vigorous seminars shaped many of the ideas found here. Among my colleagues in the institute, Joseph Ascroft, David Black, Robert Chambers, Peter Hopcraft, James Morgan, Niels Röling, Lawrence Smith, and Anthony Somerset were particularly helpful. In 1971 when I came to teach in the university's Department of Government, I was subject to another set of helpful influences. The students in my two successive classes in administrative behavior, through their own research and discussion contributions, helped me to relate my findings to organization theory and to discover similar patterns for administrative behavior in other parts of Kenyan society. I also received many helpful suggestions from my fellow teachers, Goran Hyden (now of Dar es Salaam) and W. Ouma Oyugi.

This study would not have been possible without the generous cooperation given me at all levels in the Ministry of Agriculture. The late Laxman Bhandari and Provincial Directors Gota and Kimani provided official permission for the research at different stages. My debt to the large numbers of junior and senior staff who made themselves available for our interviews and who organized their work around our demands is incalculable. Their kindness, cooperation, and patience were beyond all reasonable expectations.

Finally, I would like to acknowledge the considerable editorial assistance I received in this enterprise from my wife, Leslie. Norman Uphoff of Cornell University provided extensive, welcome suggestions for the revision of an earlier draft of this study. I also want to express my appreciation to Evelyn Martyres, Festus Gathogo Geoffrey, and Joan Allman for their typing services.

Material quoted from Peter M. Blau, *Exchange and Power in Social Life*, copyright © 1964, John Wiley & Sons, Inc., is reprinted by permission of the publisher.

Abbreviations

AA   Agricultural Assistant. A farmer contact agent, generally holding a certificate in agriculture.

AAO   Assistant Agricultural Officer. Usually a second-line supervisor in charge of a division, holding a diploma.

AFC   Agricultural Finance Corporation. A parastatal unit.

AHA   Animal Health Assistant. A farmer contact agent, generally holding a certificate in veterinary medicine.

AO   Agricultural Officer. Holds a university degree in agriculture.

DAO   District Agricultural Officer. AO third-line supervisor in charge of a district.

FTC   Farmer Training Centre. Run by the Ministry of Agriculture.

JAA   Junior Agricultural Assistant. A farmer contact agent without formally recognized technical qualifications.

JAHA   Junior Animal Health Assistant. A farmer contact agent for veterinary matters without formally recognized technical qualifications.

KTDA   Kenya Tea Development Authority. A parastatal unit responsible for all stages of small-holder tea production.

LAA   Location Agricultural Assistant. AA first-line supervisor in charge of a location.

LO   Livestock Officer. Holds a diploma in veterinary medicine, usually in charge of a division.

RO   Research Officer. Based on an agricultural research station.

PDA   Provincial Director of Agriculture. AO fourth-line supervisor in charge of a province.

PIM   Programming and Implementation Management. A Management by Objectives system.

VO   Veterinary Officer. Holds a university degree in veterinary medicine, usually in charge of a district.

Part 1   The Problem
of
Extension
Productivity

1   Asava and Masinde:
Performance at the Periphery

The Role of Government in Agricultural Development

How can developing countries force the pace of economic development in the countryside? Every day someone—an angry intellectual, an experienced civil servant, a university student, a foreign adviser—offers a new answer. Most of these solutions involve an increased role for the public services in the rural areas. Yet, another group of critics charges that the civil services of the developing world are unequal to the basic responsibility for rural development. For Africa René Dumont's dismissal is complete: "Under present conditions, there is no effective method of making a good number of the civil servants work well."[1] Henry Bienen devotes a book to arguing that neither the party nor the administration in Tanzania is strong enough to mobilize the countryside for economic development.[2] Even the governments themselves sometimes give evidence of concern "that there are serious shortcomings in the effectiveness of present operations of the Civil Service."[3] The fact that the solution to the problem of rural development is itself so often inadequate easily leads one to ask: Is there a vicious circle in development administration? Does a low level of social and economic development dictate the existence of a public administration too weak to be an instrument for accelerating development?[4]

The uncertain effectiveness of government administration in the rural areas of the developing world is gradually leading to a literature on "political penetration."[5] Interest is growing in just how governments work in the rural periphery as well as in the capital center. Even though the ineffectiveness of much of government administration in the rural areas of developing countries is now relatively well established, the potential of these public agencies for overcoming their problems and promoting rural development efficiently is, in truth, unknown. There is an urgent need for investigations of government agencies in the rural areas that not only describe their features and point out their shortcomings but also explain the dynamics of their operation and begin to

3

develop a social theory for their management. It is not enough to prove that a group of civil servants is not working well; we must know why they are not and what could lead to an improvement in their performance.

Agricultural extension organizations present an excellent opportunity to study and analyze rural administration in newly independent countries. They were established in the colonial period and have had fairly stable basic organizational patterns for a quarter century. Extension services are relatively large (Kenya's has 8,900 employees),[6] and they exist in virtually all countries. Thus it is possible to do comparative analyses of extension organizations between countries. Further, they are more concerned with the problems of increasing agricultural productivity than with the politics of distributing its benefits, so one is able to focus directly on their efficiency as an instrument for creating new wealth.

The most substantial argument for studying agricultural extension services, however, is that they are important in their own right. They are central to economic development programs as now pursued in the developing world. Thus Kenya's 1970–1974 *Development Plan* proclaims rural development as its basic strategy and calls for a one-third increase in agricultural production in five years. The extension service is one of the major parts of this program to boost agricultural efficiency.[7]

In developing countries, the state has tended to accept a fairly comprehensive set of responsibilities for farmers. At virtually every point at which the farm enterprise touches the larger economy, the governments of developing countries provide services to the cultivator. In Kenya the deep involvement of the government in agriculture originated in the colonial period. Not only did the British government conduct agricultural research and employ extension agents to give information and advice to the white farmers, it also provided extensive credit, marketing, and even sometimes labor-recruiting facilities to the settlers.[8] In the post-Independence period, the heavy dependence of large commercial white farms on government services has expanded into a reliance of all market-oriented agriculture—small and large—upon government assistance. Take, for example, the Kenya government program of increasing small farmer milk cow ownership and production. The small farmer first expresses his interest in owning a grade cow to his local Ministry of Agriculture extension worker (in many cases the agent will have encouraged him to consider acquiring one); the agent will assist him in obtaining and filling out loan forms from the parastatal Agricultural

Finance Corporation (AFC). After another extension worker, who specializes in loan applications and farm planning, has visited the farm and recommended a production plan for it, the loan application will be passed on to the AFC, which will make its own investigations and decision. Assuming that the loan is approved, an extension supervisor may be assigned to help a group of such recommended farmers find and purchase their milk cows. When the cow has been acquired, the farmer will be taught how to care for it either by a specialist extension agent or at a course at a Ministry of Agriculture Farmer Training Centre. If the cow becomes sick, it will be treated by one of the ministry's paraprofessional veterinarians. Artificial insemination will be provided at a subsidized fee (5 to 10 shillings; U.S. $.75 to $1.43) by another ministry employee.[9] Finally, the milk produced will be sold to the Kenya Cooperative Creameries at prices set by the government. Thus virtually every aspect of small-farmer milk production enjoys government-provided services. This example is perhaps extreme; government involvement in other farm enterprises, although extensive, is much less complete. The point, however, is that the present Kenyan government has assumed responsibility for a very much wider range of agricultural services than is usually the case in the Western industrialized states. In such a situation, the impact of government activities upon agricultural production can be very great.

Nevertheless, the potential importance of agricultural extension services to development does not guarantee that they always work well. For example, a study by Lever of Botswana's extension program indicates that it does not pass the cost/benefit analysis test in certain parts of the country. The value of the increased agricultural produce it creates in these areas probably is not larger than all the costs of the program (including opportunity costs).[10] Likewise, Saylor's cost/benefit analysis of the cotton extension campaigns in Sukumaland of Tanzania shows that the program probably was not a profitable development investment.[11] In a related study Saylor found only weak evidence for a positive relationship between extension agent visits to individual Sukumaland farms and their increased productivity.[12] Depressing results such as these are probably the rule rather than the exception throughout the developing world. Despite the failures, however, almost no one is willing to conclude that agricultural extension agencies should be abolished. Experiments have shown that well-run extension programs can have a dramatic and beneficial effect on farming practices.[13] Thus the emphasis of governments is on reforming their extension services to make them more productive. Decision makers rightly believe that efficiently run agricultural extension organizations would operate at strongly profitable cost/benefit ratios. Detailed analyses of extension services are needed in order to learn how such organizational efficiency can be achieved.

Such are the many reasons that convince us of the intrinsic importance of research into the administration of agricultural extension services in poor countries. This study is the result of that conviction. It is based on our work in the Western Province of Kenya between 1969 and 1972 and is supplemented by the considerable investigative and experimental research that has been done on agricultural extension in East Africa up to 1975.

The Nature of Agricultural Extension Activities

The essential role of an agricultural extension field service is that of a linking mechanism. Its classic function has been to link agricultural research centers and farmers, transmitting new technologies to the farmers and current farming problems to the researchers. More recently, in developing countries such as Kenya, extension has begun to serve as a link between the cultivator and supplies, credit, and markets. Most of the studies of agricultural extension have focused on those linkages external to the organization itself. What, for example, is the state of information transmittal and of problem feedback between the extension field service and the farmer or between the service and the agricultural research centers?[14]

Yet there is another set of linkage problems that is internal to the extension organization itself and that condition its external contacts. How, for example, is a ministry of agriculture to be organized to assure the movement of necessary information within it? Especially when government has assumed—even partially—functions played by the free market in Western economies, the administration of an agricultural extension organization becomes quite complex. As Jon Moris puts it,

> When production units are dispersed, the product mix changeable, timing important but varying between and within seasons, and where deliveries must be ordered long in advance, agricultural administration is necessarily organization-intensive. Few writers on development administration give this point adequate weight. Agricultural service activities typically consist of a chain of serially linked actions performed by various Departments and administrative actors in sequence.

A slip-up at only one or two points may nullify the impact of all preceding and subsequent actions. Even simple tasks will require a large input of organizational effort to see that they are reliably performed.[15]

No small part of the complexity of the extension administrator's task in developing nations is the management of his staff. The farmer contact agents are not full professionals and they cannot be left to set their job priorities, plan their work, or obtain the technical information they need. All of this must be organized for them by extension professionals in supervisory positions. There is a great need for research in poor countries on this personnel management function. Very little attention has been given to it or, for that matter, to subordinate staff behavior at all.[16] Yet ultimately all of extension depends upon these subordinates because they are the agents in direct contact with the farmers. In present practice they are extremely poorly managed in most developing countries. As we will show for Kenya, the result is that they do little and poor quality work.

The function of the Kenyan Ministry of Agriculture's farmer contact agents is similar to that of their Western counterparts, although their manner of performance may be different. By and large these junior staff (Agricultural Assistants and Junior Agricultural Assistants) are engaged in transmitting information from a technologically advanced source (usually a research station) to a less sophisticated potential user.[17] Sometimes this information will be conveyed in the form of advice on a particular farm's problems and potential. More often it will be part of a hierarchically approved standard package, carried in a campaign to individuals and groups of farmers. To the extent that junior extension staff are involved in providing credit, supplies, and marketing facilities, they are doing the leg work for their senior officers. They only rarely exercise any kind of managerial or decision function in these agricultural services. The heart of the agricultural junior staff role is persuasive communication. Secondarily, they are the eyes and ears of their superiors, providing the statistics and feedback needed to develop and adapt their plans, programs, and services. Only a few of the best-trained junior agricultural extension agents will make diagnoses of special farm problems, such as crop diseases or soil deficiencies. The diagnostic role, however, is more basic for the junior veterinary staff (Animal Health Assistants and Junior Animal Health Assistants), who spend much of their time identifying and treating the more common animal diseases. Whether extension agents are involved in simple campaigns or complex diagnoses, they are nonetheless the transmitters and users of information. A good member of junior staff in the Ministry of Agriculture must be skillful and diligent in receiving, understanding, using, and communicating technical information. Job descriptions for extension agents in most developing countries would be similar to this Kenyan one.

Table 1 provides a more detailed picture of the work of junior agricultural staff in Kenya. Over a five-month period in 1965–1966, E. R. Watts had nineteen Agricultural Assistants in Embu District (Eastern Province, Kenya) keep timetables of their work. These timetables were then double-checked by his enumerators. A breakdown of

TABLE 1

Time Spent on Different Duties by Agricultural Field Staff in Kakamega and Embu Districts

| | Percentage of Work Week | |
	Kakamega	Embu
Organizational tasks (staff meetings, tours with officials, courses)	15	22
Collection and processing of information for superiors	16	12
Extension of information and services to groups of farmers	20	20
Extension of information and services to individual farmers	37	46
Unaccounted for	12	. . .

SOURCES: The Kakamega data are taken from reports made in March–April 1970 by twenty-one Agricultural Assistants and Junior Agricultural Assistants assigned to locational duties in Vihiga Division, Kakamega District. These data were collected by the author.

The Embu data are based on nineteen timetables kept by nine Agricultural Assistants assigned to general duties in Embu Division of Embu District during November 1965 to March 1966. The data given here were derived and calculated from E. R. Watts, "Agricultural Extension in Embu District of Kenya," *East African Journal of Rural Development* 2 (1967): 67.

the amount of time spent on different types of work is given in the second column of the table. The table's first column is drawn from quite dissimilar data. The reports from Kakamega are unverified recollections of the previous week's work in late March and early April by Junior Agricultural Assistants as well as Agricultural Assistants in the Vihiga Division of Western Province.[18] The two data collecting exercises employed different methods of classifying the data, and in Kakamega the work of JAAs (the most junior agents) was included, while in Embu it was not. Nonetheless, the patterns of work shown are basically the same. Approximately one third of the time is spent on ministry organizational tasks and data collection. Twenty percent is given over to providing information and services to groups of farmers, and between 37 and 46 percent to individuals. Visits to individual farms take up the largest bloc of the agent's time—about twice as much as that spent with groups.

That individual farm visits are the basic method of agricultural extension in Kenya may seem incredible for a developing country. In later chapters we shall argue that extension to groups of farmers in fact would be more desirable for a variety of reasons. The individual method can be explained, however, by the favorable ratio of extension agents to farmers in most of the small farming areas of Kenya. Table 2 presents staffing figures for 1968 in a sample of fourteen small-farm districts in Kenya. In these areas the government's capacity for assistance to farmers most often is significant. For example, the average district surveyed had 308 government employees engaged in agricultural and veterinary services. It is true that many of these worked for local governments or for statutory agricultural boards, providing only specialized services. This leaves an average of 120 involved in general agricultural extension. Nonetheless, on average, a ministry agricultural agent served 558 families, with a range from 266 in Taita to 1,206 in Meru. With such a ratio, it is possible in principle for every small-farm family to be visited once a year by an extension agent. Of course this does not happen— because of an uneven distribution of junior staff, an unequal provision of visits among farmers, and the underworking of the agents. But the fact that the staff:farmer ratio is low enough to make universal farm visits even conceivable helps to explain why visits to individual farmers are the predominate method of extension. That this method does not work well in Kenya, where the conditions are relatively favorable, is one of the reasons that leads us to argue against it for other developing countries in later chapters.

A Portrait of Extension Agents

Who are the extension agents upon whom the tremendous burden of increasing agricultural production falls? In 1970 there were approximately 6,300 employees of the Kenya government engaged in agricultural extension. (An additional 2,600 agricultural personnel were doing research and the like.) These belong to four basic cadres, whose membership is largely determined by formal training in agriculture,

TABLE 2
The Density of Central and Local Government Agricultural and Veterinary Extension Staff in Fourteen Small-Farm Districts in 1968

	Agriculture Department Staff	All Government Agricultural and Veterinary Staff[a]	Number of Families	Ratio of Families: Ag. Dept. Staff	Ratio of Families: All Govt. Ag. + Vet. Staff
Central Province					
Murang'a	171	379	94,303	551	249
Nyeri	165	444	63,411	384	143
Coast Province					
Kwale	46	197	36,223	787	184
Taita	91	169	24,204	266	143
Eastern Province					
Embu	96	295	33,437	348	113
Machakos	216	460	118,145	547	257
Meru	80	443	96,494	1,206	218
Nyanza Province					
Kisii	114	384	88,346	775	230
South Nyanza[b]	173	453–	90,508	523	200–
Rift Valley Province					
Baringo	n.a.	168	32,198	...	191
Nandi	104	310	38,156	367	123
West Pokot	47	107	14,304	304	134
Western Province					
Busia	57	109	34,528	606	317
Kakamega	199	392	138,385	695	353

SOURCES: The Government staff figures are taken from J. Heyer, D. Ireri, and J. Moris, *Rural Development in Kenya* (Nairobi: East African Publishing House, 1971), pp. 44–48. The population statistics are derived from Kenya, Ministry of Finance and Economic Planning, Statistics Division, *Kenya Population Census 1969*, vol. 3 (Nairobi: Government Printer, 1971).

[a] Includes staff of local government and of the agricultural statutory boards.
[b] Local government figures are not available for South Nyanza 1968.

animal husbandry, or veterinary medicine. Using the agricultural titles, the character and relative size of these cadres is as follows: (1) Agricultural Officers (AOs), who hold a professional university degree (3 percent of all employees); (2) Assistant Agricultural Officers (AAOs), who have taken a three-year diploma after gaining the secondary school certificate (7 percent); (3) Agricultural Assistants (AAs), most of whom hold a two-year certificate in agriculture (30 percent); and (4) Junior Agricultural Assistants (JAAs), who are without any formally recognized professional training (60 percent).[19]

That the bulk of agricultural extension is actually done by 3,800 untrained agents is a fact that some Kenyan agricultural officials wish away by talk of phasing them out. Officially, recruitment to this cadre has ended, although in practice some new employment seems to occur.[20] A recent Ministry of Agriculture Working Party reluctantly acknowledged that the "basic contact officers will for the next 10-15 years be JAA's and JAHA's, who have no formal agricultural training."[21] Obviously a central problem of extension organization in Kenya is managing large numbers of untrained agents (Junior Agricultural Assistants) and of paraprofessionals (Agricultural Assistants). This management problem does not exist in the extension services of the United States and Western Europe, which use only agents at degree and diploma levels.[22] The professionals, who *are* the extension workers in the developed countries, are only the 10 percent who supervise the paraprofessional and untrained agents in Kenya. The types of personnel and the ratios between their numbers found in Kenyan agricultural extension are very similar to those found in other poor countries.[23] It is clear that even the structure and internal management problems of extension organizations in the developing countries are quite different from those of the industrialized societies.

The untrained and paraprofessional agricultural extension agents are difficult to visualize for those who have not met them. African junior staff generally have been neglected in studies of development, despite their crucial role. To make these men more concrete, there follow two personalized sketches.

The Assistant Agricultural Officer and I had finished talking about the crop extension programs he was running in Vihiga Division. As we stepped out of a red-brick divisional headquarters in Western Kenya and went across the murram-earth road toward the fields of the secondary school, we passed a jumble of sturdy black bicycles leaned against a tree and aimed for a group of nine men who were digging along a string line with heavy hoes to plant a field of corn. The bicycles we passed assured us that these were not poor agricultural laborers, but the gum boots they wore identified them as men of the soil. It was their worn khaki uniforms, trousers tucked into their boots, and safari jackets out over their belts that identified them as government agricultural extension agents laying out a fertilizer demonstration. We stopped to examine the seeds and fertilizer, and the AAO (Assistant Agricultural Officer) introduced me. I shook hands and talked briefly with them before we went on.

These men in gum boots and khaki can be seen riding their bicycles on any road in rural Kenya. They are the "change agents," the men who actually see the small farmers to persuade them to improve their agricultural practices. They are not impressive people—you may see them bent over a *jembe* (hoe) or struggling to push a bicycle through the mud—but they are important. The efforts to boost the productivity of Kenyan agriculture are ultimately in their hands.

One of the men with whom I shook hands was Asava,[24] a local man of the Maragoli subtribe of the Abaluhya. Several days later Jack Tumwa, one of my research assistants, and I called at his home. It was a Sunday afternoon, and Asava was in and glad to see us. We were invited into the sitting room and offered seats on the unpainted wooden folding chairs that recline slightly and are popular in the rural areas. The room was the largest of the three or four in the house, perhaps four by five meters. Light came in through the three glassless windows in the room's outside walls, simple catches holding open the wooden shutters. Below us was a dirt floor, above, a corrugated tin (*mabati*) roof; around, walls of dried mud on a frame of poles. The *mabati* roof distinguished Asava as among the area's most prosperous 25 percent.[25] But since the house was not made of cement or brick, it was not "permanent" and thus not an elite dwelling in local eyes. The house must have been newly built, for the walls were still not finished. The cakes of earth and the poles had not yet been covered with that good plaster of mud and cow dung, which is used in Western Kenya (and is essentially odorless).

Asava did not understand English very well. He was fluent in Swahili, a Bantu *lingua franca* extensively used in government and commerce, but we wanted to put him completely at ease and so Jack Tumwa carried the conversation in Luhya and translated for me. I had a chance to study Asava in the way one can when listening to someone without being able to understand him. He was a small, strong man of about thirty. His face and arms were quite black; obviously he was used to working in the sun. He coiled slightly forward in his chair and with a habitual intensity gathered his brows beneath his short hair. He looked the hard-working, ambitious man that our conversation and later contact with him showed him to be. His wife brought us tea and buttered bread; and as we drank, Asava relaxed to his normal level of mental and muscular tension. (The house received official visitors often enough so that the traditional meal of honor of boiled chicken and cornmeal porridge was not automatically prepared.)

We questioned Asava on a wide range of recommended agricultural practices and found that he knew his job fairly well. He quoted exactly the recommendations on how to achieve the correct plant population for hybrid corn and knew the names, prices, and quantities of application of corn fertilizers. Asava's speciality was coffee extension at that time, and we were therefore somewhat surprised that he did not mention mulching as an alternative to hoeing when he discussed the advantages to coffee output of keeping the tree area weed-free. On the other hand, he knew how to raise the milk production of unimproved cattle by improved feeding and by the use of artificial insemination to breed better strains. He was weaker on the techniques of keeping grade cows, for he had not enough money to buy any of his own. In fact, Asava was generally best informed about the crops and animals he raised on his own 1.5-hectare (3.7-acre) farm. As we saw when we left, he followed his own advice and kept a good farm with corn, beans, coffee, and one cow. Although his work commitments forced him to use hired labor, his farm was only marginally larger than the average for the area (0.8 hectares; 2 acres) and was a good example to his neighbors. Asava did not know much about soil types, plant nutrients, and the other technical aspects of his trade, and consequently he could not offer very profound explanations of why his recommendations worked. But he knew from personal experience what did improve productivity and profits, and this must have made him an effective change agent in his area.

Asava's job was to know Ministry of Agriculture recommendations on how to improve the productivity of small farms and to convince people to adopt these innovations. He did this effectively, if simply, and was faithful and diligent in his work. When we asked about the work he had done in the previous week, he consulted his pocket diary. He had spent Monday and Tuesday laboring on and conducting a demonstration attended by thirty farmers; Wednesday he had visited various demonstration plots with the District Agricultural Officer (DAO); Thursday and Friday had found him on six different farms, instructing their owners in the spraying, pruning, or fertilizing of their coffee trees; Saturday he had been at the cooperative factory supervising coffee bean grading and processing. The last day's work was evidence of particular diligence, for official policy was to give extension agents Saturday off to work on their own farms. In the performance of all these duties, Asava traveled by bicycle. He told us that he would pedal for an hour and a half to reach the furthest farmer for which he was responsible. But it took him only half an hour to get to divisional headquarters for official meetings, for he lived relatively nearby. Some of his colleagues needed two or three hours for such trips.

Asava's official identity was that of Junior Agricultural Assistant assigned to coffee extension for his location (an administrative unit, of 92 square kilometers in this case). His rank was a function of his formal education and his technical training. He had spent eight years in school and had earned the Certificate of Primary Education. With no secondary education, Asava had never had the chance of direct admission into any of Kenya's training institutions for extension agents. Instead he had been taken fourteen years earlier into the "untrained" category of Junior Agricultural Assistants. Each year he spent one week at a residential training course taught at a local Farmer Training Centre by ministry officials with degrees and diplomas. He had also spent a week on a coffee course before becoming a specialist on that crop. Despite two small promotions, considerable field experience, and a cumulative total of nearly four months of residential instruction, Asava was still officially classified as untrained. In fact, he seemed to fear that if the ministry realized how much he had forgotten of his primary school English, he would be considered "untrainable." [26] A few months later, we found him on a ministry training course for those fluent in English. Asava would never pass up an opportunity that might lead to a promotion.

The junior staff of Kenya's agricultural extension services are of many types, and Asava is one of the better. Further insight into the ministry's organizational problems can be gained from an example on the other extreme. Masinde is employed as an Agricultural Assistant (AA) in Marama Division of Western Province. He has been with the ministry for over thirty years and is about to reach the mandatory retirement age. He is a relatively well-to-do man. He has 2.7 hectares (7 acres) of land, fifty coffee trees, and ten cattle. His house has a *mabati* (corrugated iron) roof and the walls are good and solid; it would be called a semipermanent dwelling. Masinde's prosperity is best shown by his ability to pay school fees for his many children, one of whom is now in a university overseas.

Humphries W'Opindi, another of my research assistants, comes from the same general area as Masinde. W'Opindi, himself the son of a former Junior Agricultural Assistant, is now a graduate in agriculture from the University of Nairobi. His recollections of Masinde are vivid.

"Masinde gave me a very poor impression of agricultural extension workers. When I was in upper primary school, I came to realize more and more that he and many others were just enjoying themselves in the ministry. This was not so evident in the late '50's when I was younger. There was then a lot of noise about building terraces, and I used to meet Masinde early in the mornings hurrying to supervise the work. A colonial officer had once slapped him when he could not explain his work, and he was anxious to avoid their anger. In 1962 as Uhuru [Independence] approached, Masinde suddenly stopped going to work early in the mornings. I never met him any more and instead saw him passing by my school around 10 A.M. At lunchtime he was on his way home again! This became worse and worse so that by 1964 he would not report for duty at all one or two days a week. He used to spend most of his time at home ordering his children to move this or that cow from tree to tree.

"Masinde knows very little about agriculture. When he was first employed he went to Bukura for a six months' course, which in those days qualified him to be an Agricultural Assistant. I'm told that he got into the course only because one of the trainers was his relative. His knowledge and that of a typical Kenyan peasant are not very different. He knows only things like: "Manure is good for crops." "It is good to plant in straight lines." "Grade cattle are always better than local ones." If a farmer were to ask him to explain, he would definitely be at sea.

"Because he knows so little, he concentrates on visiting poor farmers and his friends. After drinking a cup of the farmer's tea, Masinde either praises his farming or turns to a non-agricultural subject. He is really useful only when a farmer wants to get something from the Ministry. He can recommend you for free cotton seed, or a Government loan, or permission to plant coffee. Friends and those who serve tea profit accordingly. Lying is cheap when it comes to writing reports about one's work. (I used to write false reports for my father.) The Location Agricultural Assistant (LAA) who is supposed to supervise Masinde and the other junior staff will not report them unless they do something grossly wrong. The LAA has come up through the Ministry with his subordinates; they are his friends and he would not like to quarrel with them.

"Masinde does very little on his farm. He uses no fertilizer on his corn, yet he advises others to do so. An unemployed villager wondered to me one day why Masinde tells people to do what he himself does not do when he is the one with the salary and can certainly afford it. He is such a poor farmer that more than once he has been hit with hunger and had to borrow corn from other farmers. When coffee was first introduced his care for it helped to scare some of the enthusiastic farmers away from it. He was one of the early ones to plant coffee and began by husbanding it carefully. Then suddenly he began to ignore it, let it go bushy, and tied his cattle among the trees to graze. The local opinion was that if coffee had proved too much for a salaried man like Masinde, then they too would be unable to manage it."

Humphries W'Opindi's account gives a graphic picture of bad extension work and of an agent whom the ministry will be delighted to retire. These personalized descriptions of Asava and Masinde have been long in order to convey a real feeling for the material with which the ministry is working and the problems and prospects it faces. With this picture in mind the following generalized and theoretical analysis may be more real.

Development Agencies and Organization Theory

How are we to attempt to understand the extension agents in Kenya's Ministry of Agriculture and the problems of managing them? When one is confronted with the real world of extension in Western Province—with Asava and Masinde, the agents; Makhanu and Kinyanjui, their supervisors; Khamadi and Kikaya, the farmers—there is a tremendous temptation to present them as explanations in themselves. Individuals everywhere have a uniqueness that defies the generalizations of the social scientists, and those met in a culture other than our own have an even more fragile particularity that is easily broken with facile talk of types and models of behavior. Yet, to succumb to the beauty of the particular is to confine one's knowledge to the anecdotal and to limit one's ability to understand other unique but related individuals and human situations. If we can use social theory to explain one portion of a person's behavior without simultaneously destroying its individual integrity, we will understand both him and others like him better.

Organization theory immediately suggests itself when we are looking for ways to understand and manipulate bureaucracies. Within its confines exist systematic and empirically based propositions on supervision, staff motivation, feedback, managerial structures for innovation, and so on—all centrally important to extension administration. But very little research has been done on the applicability of organization theory to non-Western cultural settings.[27] Certainly, since the publication of Michel Crozier's *The Bureaucratic Phenomenon*, it is impossible to assume a priori that culture makes no difference to the fundamental human relationships in bureaucracies. Crozier argues convincingly that certain aspects of organizational behavior have changed meanings in different national settings. For example, he suggests that differences in the French and American conceptions of freedom lead to different types of bureaucratic behavior in the two countries.[28] The applicability of the sociology of organizations literature to an African development agency is also clouded by its almost complete lack of studies in rural settings, even in the West. Since evidence does exist that a formal organization is powerfully affected by its environment, the usefulness of Western organization theory in rural Africa must be seriously examined.[29] For example, Goran Hyden's work on Kenyan cooperatives shows that distribution politics so dominate the committees of these organizations that theories which assume an interest in efficiency will not fit.[30]

Nonetheless, fundamental differences between the rural areas of poor countries and the urban West do not mean that they have nothing in common. Power and authority, for example, may take different forms in various societies, but they do exist throughout. To the extent that a given theory is dependent on the existence of authority and not any particular form of it, that theory will be useful in all cultural settings. We will demonstrate that such cross-cultural theories do exist. Colin Leys and Patricia Stamp found American theories on organizational goal adaptation applicable to their study of the early life of administrative training institutions in Kenya.[31] Even Michel Crozier—an advocate of cultural relativity—has developed general theories on power that were found to work well in Silas Ita's study of Kenyan chiefs.[32] Similarly, bureaucratic organizations are universal in today's world and their distinctive authority systems tend to produce common patterns of behavior in all cultural settings. Our problem then is not to accept or reject organizational theory as a whole, but to sort out which of its propositions transcend their industrial Western origins. In this book we indicate the ways in which such cross-cultural theories can be modified and applied in developing countries to facilitate our understanding and practical management of their extension organizations.

Organization theory covers a vast range of problems, from the survival struggles of large bureaucracies to the problem-solving behavior of four-man experimental groups. No single study could deal with such an array of problems. The set of questions into which we are seeking insight concerns the work behavior of subordinate staff. A very large amount of research has been done on the determinants of worker productivity in the West, and organization theory in this area is well developed.[33] The understanding of these social processes in what Amitai Etzioni calls utilitarian organizations is even further advanced. A utilitarian organization is one in which the leadership exercises a basically remunerative (wage-based) power over its subordinates and the involvement of lower participants is based on a calculated commitment (rather than a coerced or moral one). Most government agencies and businesses are of the remunerative type and industrial sociologists have devoted a great deal of study to them.[34] As it happens, however, almost no systematic studies have been carried out on worker productivity in utilitarian organizations in Black Africa.[35] For both practical and theoretical reasons, we feel it to be imperative that this vacuum be filled. In doing so, we would gain important insights into the determinants of the morale of African employees and of their relationships with their superiors and fellow workers. Our understanding of a whole range of crucial theoretical concepts hinges on continuing research in this area: formal and informal organization, vertical and horizontal communication, the nature of authority, and so on. The practical consequences of such insights are also likely to be great. From them can come more efficient and hence more beneficial rural development agencies.

Before turning to the details and analyses of our study of Kenya's extension services, we would like to explain our position on one of the major organization theory debates. We have stressed several times already our interest in improving the efficiency of the extension service in general and the quality of work of its agents in particular. Such a focus would appear to leave us vulnerable to one part of what has been called the Structuralist critique. Amitai Etzioni, for example, has faulted the work of many industrial sociologists because they ignored the fundamental difference in interests among many of the competing groups in organizations. In concentrating on improving the goal achievement of the organization, they implicitly favored the position of management against the workers.[36] We agree with the Structuralist position in general, but we do not accept such a criticism of this study.

We acknowledge that our analysis is made from the perspective of one of the interest groups concerned with the extension organization. That viewpoint is neither that of the extension management nor of its workers, however, but of the mass of small farmers that it is supposed to serve. Through direct and indirect taxation, Kenya's farmers pay for the provision of agricultural services. As the cost/benefit studies cited above show, the value of the services they receive is often less than has been paid for them. Kenya's small farmers are not starving, but very few of them earn incomes better than those of the most junior civil servants. Thus it seems to us that inefficient agriculture bureaucracy is more exploitative of the farmers than the extension management might be of their workers.

This is not to say that society's interests in efficient government always should override the desires of civil servants for better incomes or a relaxed work pace. When government employees have lower incomes or poorer working conditions than does the average citizen, they may claim a moral right to demand a better return for their services from society. But in Africa even the lowest paid civil servant is a member of a labor aristocracy when his lot is compared with that of the great majority. Also, when most people in a country are subordinates in one or more formal organizations, there may be an implicit agreement in the society to allow some of the potential income from those organizations to be consumed by the workers, before it is produced, in more human and inefficient methods of work. But in countries such as Kenya, most citizens are not wage earners and inefficiency in the nation's productive organizations is an implicit tax upon the masses, which brings them no benefits. As the socialist president of Tanzania has written, inefficient bureaucracy can be as exploitative of the masses as is capitalism.[37]

Our basic value position in this study therefore is in favor of the small farmer and greater efficiency in the extension organization. We are unapologetic about this, but nonetheless we must recognize that the extension agents and agriculture administrators have interests of their own to defend. In chapter 12, we will deal directly with how far the interests of the ministry's management are compatible with extension efficiency. There we will see that the achievement of efficiency would be more an act of charity than of self-interest for the higher echelons of the civil service and government and that programs to achieve efficiency can be an imposition upon this managerial group. The competing interests of the subordinate extension staff will be evident throughout our study. Because these workers are actually in a fairly strong power position, however, we will show that they are able to demand better material terms of service in return for the cooperation that would be necessary to achieve greater efficiency. As the extension workers seem morally ill-at-ease with the leisure they enjoy at the expense of the farmer and as they seem happy to trade such leisure for the better career prospects we recommend, we do not feel that our study jeopardizes their true interests. In fact, in the policy debates in Kenya into which this study has thrust us, we have found ourselves to be among the strongest advocates of agents' interests, particularly those of the less educated ones.[38]

To say more on this issue here would be to anticipate the analyses and conclusions of our study. We hope that we have said enough to give a sense of the values which we have brought to our work and thereby to enable the reader to judge the relevance of our conclusions to his own views and interests.

2 The Measurement of Extension Work

The Theoretical Context of Measurement

Kenya's agricultural extension services are not effective enough. Why? The poor quality of farmer contact agents has become a commonplace answer, not only for Kenya but also in the rest of the developing world.[1] Yet, this poor quality is usually attested by anecdotal evidence only and rarely is it examined closely enough for serious reform measures to be undertaken. We need to measure carefully the performance of junior extension staff in order to identify the extent and nature of the problem. Before we turn to our collection of evidence on agent capabilities and work, however, we would like to stress that the need for measures of performance is not peculiar to this study. It is central in all administrative research.

In any study of an administrative organization, the basic factor to be explained is output. The primary interest of political scientists and other citizens in government organizations usually is in what they produce for the system.[2] By and large, our concern with issues such as political control of government bureaucracies, corruption, and patterns of recruitment for the civil service is based on the presumed effect of these factors on the creation and distribution of public benefits. The greatest number of Fred Riggs' propositions about comparative administration, for example, ultimately relate to the productivity of the civil service.[3] The concept of productivity must be broadly conceived in administrative studies, of course, and should include consideration not only of how much benefit is produced but also of who gets it. Development studies have tended to be dominated by fairly narrow considerations of economic growth or productivity. There is great need for sensitivity to noneconomic benefits and to the patterns of distribution of benefits as well—not only to those outside the organization but also to those within it.

Our conviction that the issue of output is central to the study of administration is so strong that we would make it a basic methodological principle. We believe that all studies of public administration should incorporate some measurement of outputs. In our experience a reasonably objective indication of what an organization is actually producing and the effects it is having on its environment puts the other organizational characteristics into clear perspective and focuses attention on the truly critical issues. Otherwise administrative researchers tend to treat as ends in themselves factors that are only presumed to contribute to particular types of outputs. In the process sight is lost of the relative importance of these factors in producing that output. From a more practical point of view, we know of no stronger impetus to administrative reform than clear, objective evidence of poor performance in a significant output area.

A recent Kenya Ministry of Agriculture Working Party report contains the following statement on the aims of extension service:

> The agricultural extension services should aim to give the farmer advice and services which enable him to run his farm and home business more successfully. In its orientation, the extension service should cover the whole range of farmers from the best to the poorest.[4]

We knew that these aims were not being adequately met before we began our research. Statistics on the rates of adoption of farming innovations, such as those cited in the Preface, indicate that the effectiveness of agricultural extension needs to be improved. A wide range of suggestions has been made about the causes of extension ineffectiveness in developing countries. Producer prices in many instances are too low to be a real incentive to the farmer. There are doubts that some of the farming innovations are actually profitable to the small farmer when labor costs and competing opportunities are properly calculated. Investment capital and, hence, credit shortages are often cited as constraints on small-farm production. Inadequate arrangements for the convenient and timely supply of inputs and marketing of produce are also identified as frequent problems. Existing methods of teaching farmers about new husbandry techniques leave much to be desired. All of these problems are real and important. Much is already known and has been written about them. Furthermore, most of these problems are less acute in Kenya than in the majority of developing countries. The one area which we feel has received inadequate attention is a whole range of problems concerning the motivation and management of farmer contact agents in developing countries. The major empirical contribution of our work is to fill the gaps in our knowledge in the area of subordinate staff management, although our findings have much broader implications. Especially from chapter 8 forward, we use our insights into the lower echelons of the Ministry of Agriculture as a basis for analyzing and making suggestions about problems of extension organization at district and national levels.

Despite our broad interests, our own field research work concerns the contribution of the subordinate staff of the Ministry of Agriculture to the achievement of its official goal. The output of the junior agricultural extension agents therefore is the basic dependent variable in our analysis. As we said earlier, the way in which this output is to be measured deserves very careful attention. The choice of an indicator for any one variable in a projected study cannot be made in the abstract. To be most useful, the measure selected has to be consistent with the purpose of the research. Any one indicator should be easy to relate to the others in which we are interested. To achieve such internal consistency we must give thought to the full range of variables and the units of analysis with which we will want to work.[5]

The existing body of theory on the utilitarian type of organization leads us to expect particular sets of variables to have important effects on subordinate staff output. We will deal in detail with these theories in the course of the substantive chapters which follow, yet it will help to explain the methodology we have chosen if we review here the kinds of questions they pose.

One set of variables concerns the motivation of the individual extension agent. Portions of his personal background can be expected to shape the way that he responds to farmers, his work, and the organization and the incentives it offers. For example, social class, education, professional training, and length of service in the ministry all may be expected to socialize the individual to certain expectations and modes of behavior. Do they actually affect individual agents, and if so, how? What other social characteristics have a bearing on the way that extension staff do their work? Data about social traits such as these could be collected most easily in a sample survey of extension agents.

A second set of variables concerns the authority system of the organization. How effective are the agents' immediate supervisors? Do they rely exclusively on their formal sanctioning powers or do they also exercise leadership? How do the junior staff respond to formal authority? What characteristics of supervisors make them more effective leaders? To answer these questions, information about supervisors and the groups of junior staff working under them clearly is required.

A further group of questions involves the peer group organization of the extension agents. Are there informal ties among the staff? How strong are they and how do they operate? How do the goals of any informal organization of agents relate to the goals of the ministry? What kinds of conflicts exist between the extension workers and the ministry's field management? To deal with issues such as these we also will need data about groups of staff rather than just individuals.

The final set of issues concerns the structures used by the organization to achieve different types of work behavior. What incentives does the ministry provide for its employees? How does it plan and evaluate work? By what means is information transmitted within the organization? What is the effect of these structures on the output of the extension workers? Answers to these sorts of questions ideally would derive from comparisons between different whole extension organizations, but lack of money and time prevented such an anaylsis. Instead, we sought to isolate functional sections and individuals within the ministry that had differential exposure to these structures and to compare the impact of the structures on them.

The theoretical and practical questions that we have raised indicate three units of analysis for our study: individual extension agents, work groups of junior staff, and functional sections of the ministry (the last unit acting as a partial substitute for analysis of the whole organization). In order to work with these three units of analysis simultaneously, we drew a sample of geographical areas and studied all the agents, supervisors, work groups, and functional sections working within them. (More will be said about our sampling strategy later.) Information about the explanatory variables that interested us for each of our three basic units of analysis was most easily learned through interviews with the individuals who fell within our sampling frame. Thus, at least for our independent variables, our method of data collection was to be survey interviews.

Having defined those aspects of extension performance that interested us, identified the units of analysis with which we were to work, and decided on the type of data that would be best for learning about our independent variables, we had to choose how we would measure the output of the junior staff. Our experience in the field indicated that we would be wise to use partial indicators of organizational performance directly related to the work process in our basic units of analysis. We came to realize that attempts to be thorough in measuring the degree to which the extension organization was achieving its goals would actually hinder our efforts to learn about the determinants of the agents' work performance.

The dangers that comprehensive measurements of goal achievement pose for our type of study can be illustrated by a brief discussion of the two best methods for making an assessment of overall extension performance. The first is cost/benefit analysis, in which all the costs of a project are calculated and then compared with the monetary value of the increases in crop production that have been achieved. In the conditions of developing countries, the margins of error that must be accepted in the resulting profitability ratios are quite large.[6] But from our point of view, lack of precision was not the major problem. Rather it was that cost/benefit ratios are appropriate to study at levels of analysis and with types of questions with which we were not dealing. Cost/benefit analysis measures the productivity of a project as a whole and thus gauges the final impact of all the possible influences on the project. In a cotton project, for example, these would include farmer responses, soil character, international prices, the quality of agricultural research, the cotton marketing organization, and the agricultural extension organization, not to speak of other government bodies that might be called upon to participate. If one were interested in assessing the productive consequences of differences in extension organization, it would be necessary to sort through a large number of other, non-administrative, causal variables before getting down to the impact of one's unit of interest. Once there, one would be dealing with the extension organization as a whole and would be unable to probe for differences in productivity within it. Cost/benefit analysis would not help us in deciding whether and in what way the junior staff were responsible for any failures in goal achievement.

A different set of problems would arise if one were to attempt to measure the productivity of the extension service by interviewing farmers regarding how much impact the agents had had on their farming. When we first began our research, we planned to study the variations in junior staff productivity by interviewing a random sample of eight farmers for each of a sample of extension workers, and we actually pretested this method.[7] The logic was correct, for we had identified the extension agent as the real unit of analysis and farmers were being used only to provide an estimate of agent effectiveness. In practice, however, such a measurement strategy would have led to great problems. First, it would have been prohibitively expensive and time-consuming. We would have needed to interview a random sample of 1,600 farmers in order to make a rough assessment of the effectiveness of 200 agents. Second, the analysis of the farmer interviews would have taken on a life of its own, directing our intellectual resources and hypotheses away from the units we wanted to study and toward the problems of the farmers.[8]

Because of the pitfalls of comprehensive measurement of productivity, we decided that we should work with partial measures of output that were fairly simple to make and were as close as possible to the level and type of organizational behavior in which we were interested. We did accept, however, that precisely because these measures were partial we should use several different ones and give particular attention to their reliability. We felt that a plurality of indicators was especially important because different aspects of work performance are likely to be unrelated to one another and may be quite unalike in their relationships to important explanatory variables.[9]

Finally, there was the question of the type of data we would use to measure work output. Since the other information needed was best obtained through interviews with a sample of extension agents, we saw that it would be most efficient if we were able to learn about their work performance through the same instrument. Of course, it is extraordinarily difficult to get people to be honest when they are asked to report on the amount and quality of the work that they themselves are doing. We recognized that we would have to construct our questions and check the answers carefully if we were to get valid and reliable data.

The Indicators of Extension Agent Output

In chapter 1 we observed that the basic function of junior staff is the persuasive communication of technical information, primarily to individual farmers. Secondarily, junior staff have some responsibility for contributing to the process of adapting technical recommendations and programs to local conditions. We have already noted that it was not feasible to measure directly the extent to which extension agents were contributing to increases in farm income in their areas. Nonetheless, we do know several things that would enable us to infer from partial indicators of performance the kind of contribution being made by an agent to farmer welfare in his area. First, thanks to a relatively old, large, and well-endowed agricultural research establishment, the Ministry of Agriculture does have a large number of technological innovations to suggest that do represent a profitable improvement on the way that farming is being done on the majority of small farms in Kenya.[10] Second, the agricultural extension services are the primary source of information about these innovations for small farmers.[11] Third, many farmers are adopting these innovations, and the evidence suggests that their propensity to do so is increased by their direct contact with an extension agent.[12] We therefore can assume that a visit to a small farmer by a well-informed agent which led to the persuasive communication of some one of the official recommendations would generally be a beneficial thing for the farmer.

Nonetheless, we have also stated that it was not possible to measure directly the amount of persuasive communication achieved by each agent. We do know what the necessary preconditions for beneficial persuasive communication are, however, and some of these preconditions can be easily measured. First, the agent must know the information that he is supposed to be communicating; if he does not, he cannot pass it on to the farmer. Second, the agent must be able to explain the innovation and its value to the farmer in order to persuade him to adopt it; if he does not understand what lies behind his recommendation and what its benefits are, he is unlikely to change many minds. Third, in an extension system based primarily on visits to individual farmers, an agent must make visits in order to communicate; an extension worker who sees few farmers cannot have many converts. If we had information about an agent's technical knowledge, explanatory ability, and number of farm visits, we would be able to make a reasonable inference about his amount of persuasive communication with farmers and the degree to which he was helping to increase farm incomes in his area. A similar process of measurement by inference can be applied to other aspects of the extension worker's job. This was the logic that led us to the performance indicators which we used in this study.

Agricultural Informedness

Since, as we have just argued, an extension agent's work is communication, a necessary condition of his success is knowledge of the technical information he is supposed to transmit. We therefore devised a test to assess this part of the junior staff member's competence. The test was administered orally and in the vernacular to keep the measurement of technical informedness uncontaminated by literacy and skills in English or Swahili. Each individual was asked four cycles of questions in four of five different technical areas: fertilizer applications for hybrid corn, feeding programs for grade milk cows, the diagnosis of east coast fever in cattle, the identification and prevention of berry borer infection in coffee, and the identification of the spiny bollworm in cotton. All of these topics are important to the agricultural extension program in Western Province. They were selected, the questions drafted, and the answers determined only after careful consultation with various national specialists.[13] The questions on east coast fever were designed primarily for veterinary staff, although they also were asked of agricultural agents to test their general knowledge. The questions on coffee and cotton were used alternately, depending on which of the two was being grown in the area where the agent was working.

The performance on these objective information tests indicated some major shortcomings among the junior staff. The agents were expected by their specialist superiors to know all the pieces of technical advice about which we asked and to be disseminating them to farmers. On average, the agricultural workers knew 72 percent of the requisite pieces of information on corn fertilizers, which is just satisfactory, but they knew only 42 percent of the information on berry borer in coffee, far below the expectations of the national coffee experts. The junior veterinary staff averaged 59 percent of the information on east coast fever, the diagnosis of which is basic to their work.

As can be seen from table 3, the ability to answer a set of technical questions on one topic is correlated with ability in other areas for agricultural agents. As would be expected, the veterinary questions on east coast fever are more marginally related. It is clear that there is a basic factor that can be labeled *Agricultural Informedness* and can be measured by summing the individual's scores on the corn, milk cow, coffee, and cotton questions. This variable is one of our basic measures of junior staff productivity. The major analyses of this aspect of performance will be made in chapters 6 and 7.

Explanatory Ability

Our preliminary measures of Agricultural Informedness were pretested on a small sample of extension agents in March 1970. Our analysis of the results of this pretest led us to fear that the indicators used—objective information tests—might be producing superficial results. The tests concentrate on the ability to recall important details, but an extension agent must also be able to understand the practices he is recommending, explain them, and persuade farmers to adopt them.[14] To measure such in-depth grasp of agriculture, we decided to follow each cycle of

TABLE 3

Intercorrelations among the Degrees of Informedness of Junior Agricultural Staff

	1	2	3	4	5
1. Hybrid corn fertilizers	X				
2. Cotton pests	.49	X			
3. Coffee pests	.23	...	X		
4. Milk cow feeding	.63	.56	.35	X	
5. Cattle disease	.25	.24	.10[a]	.25	X

NOTES: All correlations with item 2 have an N = 97. All those with item 3 have an N = 72. All other correlations have an N = 169.

The statistic reported here is a correlation coefficent. It measures the strength of the relationship between two variables and will range from 0.00 to 1.00, depending on how closely related the two are. All correlations given here are positive unless preceded by a minus sign. Correlation coefficients are all comparable with one another, the higher number indicating the stronger relationship. When a correlation coefficient appears by itself, it is signified by an r (e.g., r = .63). The following guidelines for assessing the strength of a relationship are generally agreed:

greater than .70　High correlation: marked relationship
.40 to .70　Moderate correlation: substantial relationship
.20 to .40　Low correlation: a definite but small relationship
less than .20　Slight correlation: a relationship so small as to be almost negligible

(See T. G. Connolly and W. Sluckin, *An Introduction to Statistics for the Social Sciences*, 3d ed. [London: Macmillan & Co., 1971],p. 154.)

[a]p = .20. All other correlations are statistically significant at the .05 level. The p refers to the probability that the relationship suggested by the data actually does not exist and has appeared only as a result of the chance selection of the sample. The p then gives the odds that we are in error. By convention we usually work only with relationships that are statistically significant at a level of $p \leqslant .05$; in other words, there is no more than a 5 percent chance that we are in error.

technical questions with another query: "Going back over what we've just discussed, could you please explain to me why what you see as best is a desirable practice? Exactly what does it do agriculturally speaking and how does it increase yields?" The interviewer then recorded a subjective rating of (1) the respondent's understanding of the technical processes involved and (2) his knowledge of yield differentials. These two ratings are so closely intercorrelated that we have combined them for each technical area and call the result the agent's *Explanatory Ability* on that subject. As can be seen from table 4, Explanatory Ability in an area is generally closely related to Informedness on the same topic (see line 6). The only exception is hybrid corn fertilizers, where the level of Informedness is relatively high and all the respondents had a fair amount of direct experience with the crop. Table 4 also demonstrates that, unlike Informedness, Explanatory Ability on one subject does not tend to be correlated with Ability in others. In-depth understanding of one technical topic is not necessarily accompanied by a good grasp of other areas. At this level, some specialization of skills is evident. Nonetheless, in contrast to Informedness, the general level of Explanatory Ability is fairly high. For each of the various agricultural areas tested, almost half of the extension agents were rated by our interviewers as being high in their understanding and as having a good idea of the yield consequences. We will return to a detailed analysis of Explanatory Ability and Informedness in chapter 7.

TABLE 4
Intercorrelations of Explanatory Ability of Junior Agricultural Staff

	1	2	3	4	5	6
1. Corn fertilizer explanation	X					
2. Cotton pest explanation	.19[a]	X				
3. Coffee pest explanation	.02	. . .	X			
4. Cow feeding explanation	−.02	.14	.26[a]	X		
5. Cattle disease explanation	−.06	.36[a]	.11	.33[a]	X	
6. The Respective Information Test[b]	.07	.44[a]	.98[a]	.74[a]	.62[a]	X

[a] $p < .05$.
[b] Line 6 represents the correlations of Explanatory Ability on one particular topic with Informedness on that same topic.

Innovativeness

Kenya has a very wide range of ecological conditions and soil types. Obviously a wide range of agricultural and animal husbandry practices need to be varied with local conditions in order to achieve maximum yields and profits. The appropriate fertilizer mix for corn changes with the soil type, grade cows do not need as much pasture in heavy rainfall areas, and so on. Yet technical advice for Kenyan agriculture is most often given in one or two national packages and adaptation to local conditions is formally left to the field extension officer—the AAO or AO. An important criterion of a fully effective extension service is therefore the extent to which it is engaged in making intelligent adaptations. Junior staff could participate in this innovative process either by making adaptations themselves or by bringing to the attention of their superiors the problems that indicate a need for changes in the recommendations.

In order to get at this aspect of extension, we not only asked our respondents to give us the standard recommendations for the various technical topics covered but also any adaptations that they had made in them. The Informedness scores are based on the agents' knowledge of the standard recommendations, but we also kept separate track of those staff who were at all adaptive, either completely on their own or in consultation with their superiors. We call this variable *Innovativeness*. Only 13.4 percent of the staff interviewed indicated that they had made any adaptations in any of the technical areas we covered. The particular adaptations made were not very startling ones, by and large. In chapter 8 we will discuss the reasons that the Kenyan extension services are not as innovative as they need to be.

Visit Effort

We have noted that the work of the Ministry of Agriculture extension agents consists primarily of visiting individual farmers and providing them with information and services. Our interviews revealed that junior staff spend an average of 2.9 out of 5 workdays a week on this activity. The other aspects of their work vary considerably with the season and the agent's specialization. A coffee specialist will spend part of his time at the cooperative factory grading the coffee crop; the general extension worker may be called to a location *baraza* (public meeting) or be assigned to collect agricultural statistics. It would be extemely difficult to measure how hard the agent is working at these many tasks, but we can determine the amount of effort he is putting into his basic function of seeing farmers. In order to take into account the varying functional mixes of the work of differing extension agents, we have used as our measure of work effort the average number of farmers that an agent visits on a day devoted wholly to that activity. To do this, we asked each extension worker to specify for us the days in the previous week that he spent visiting farmers and to name for us the specific farmers he saw. We thus can divide the total number of farmers seen by the number of days used for the activity and achieve a standard measure of work effort.[15] This variable is labeled *Visit Effort*.

The information obtained about the farmers visited was detailed enough to enable us to check the validity of some of the more suspicious entries. Somewhat less than a quarter of these nonrandom checks uncovered deception on the part of the extension worker. No deception was discovered for those interviewed by one of our three interviewers, however. Statistical analysis shows that the responses achieved by this interviewer are quite different on this item but not so on other potential problem variables. From this we conclude that the responses he gained for this question are honest, and we have therefore adjusted the responses received by the other interviewers downward.[16] Having done so, we conclude that extension agents visit an average of 1.75 farmers on a day that is devoted wholly to that activity and that they see a mean of 20.3 farmers a month. These statistics indicate quite a low level of Visit Effort by the extension staff. This disappointing picture is corroborated by the general observations of ministry extension officers—AAOs and AOs—and by research done in similar conditions in Tanzania.[17]

Dishonest reporting obviously presents major problems in analyzing the factors associated with Visit Effort. Does high Visit Effort represent harder-than-usual work or particularly audacious deception? In practice the problems have proved surmountable. As will be seen in chapter 6, the factors associated with high Visit Effort in individual extension workers are related to other indicators of agent performance in much the same way. This gives us confidence that Visit Effort is, however imperfectly, basically measuring performance rather than deception. In chapters 4 and 5, Visit Effort is studied by itself, but there we examine the average Visit Effort of groups of extension agents and not that of individuals. We feel confident that the adjustments we mentioned above have given us a fairly accurate measure of Visit Effort at the group level, where individual differences in the tendency to be dishonest are averaged out.

Progressiveness Skew

Our final indicator of the output of the agricultural extension agents concerns their distribution of benefits among their clients, the farmers. The Kenyan government prescribes that the visits of agents to individual farmers should be made free of charge. These visits thus constitute a gain for the recipient and the official government policy is that "the whole range of farmers from the best to the poorest" should benefit by them to at least some degree.[18] It therefore is important to ask just how they are actually being distributed. To probe this element of performance, we asked each agent to tell us not only the names of farmers he had visited in the previous week but also something about each one. From these descriptions we are able clearly to identify the farmer as being progressive, middle, or noninnovative. The proportion of an agent's visits made to farmers in each of these three categories can then be calculated. We have just noted that some agents falsely inflated their farmer visit numbers. We compared the average percentages in each of the three farmer categories that were derived from the interviews conducted by Luchemo, where there was no deception, with those done by Tumwa and W'Opindi, where there was. We found no significant differences. We concluded that agents who were reporting more visits than they actually made were accurately indicating the farmers they would have seen if they had done more work. The percentages thus seem to be fairly reliable estimates of the actual distribution of extension services. Our figures indicate that the provision of extension visits is greatly skewed in favor of the progressive farmers. As we will show in chapter 9, 57 percent of the visits are made to the 10 percent of the farmers who are progressive, compared with 6 percent of the visits to the 47 percent who are noninnovative. We label this variable the *Progressiveness Skew*.

In summary, this study is based on five partial indicators of the output of junior agricultural extension agents. These indicators are directly related to their day-to-day work, and we believe that they are reasonably reliable. We have developed measures of Agricultural Informedness and Explanatory Ability in order to assess the extent to which the agents know and understand the recommendations they are supposed to disseminate. Visit Effort is an indicator of how hard they are trying to disseminate these recommendations. We also have measured the agent's Progressiveness Skew in order to determine the distribution of his services among classes of farmers. Finally, we have noted how far the extension worker gives evidence of Innovativeness and therefore is creatively adapting standard recommendations to local conditions and providing his superiors with the feedback they need to do the same. Together these five partial indicators give us a good overall picture of junior staff performance. With them we can make a detailed analysis of the organizational determinates of effective extension work.

The Research Site and Sample

In order to complete the methodological background for this study, we now need only to describe our research site and sampling strategy. Our surveys were conducted in Kenya's Western Province during March, July, and August 1970, March and April 1971, and June 1972. We chose Western Province because of the personal contacts we had there. Initially we had planned to extend our study to other areas of Kenya as well, but as we went into greater and greater depth in Western Province this became impossible. (See map 1 for the location of Western Province in Kenya.) Nonetheless, we have strong reasons for believing that the

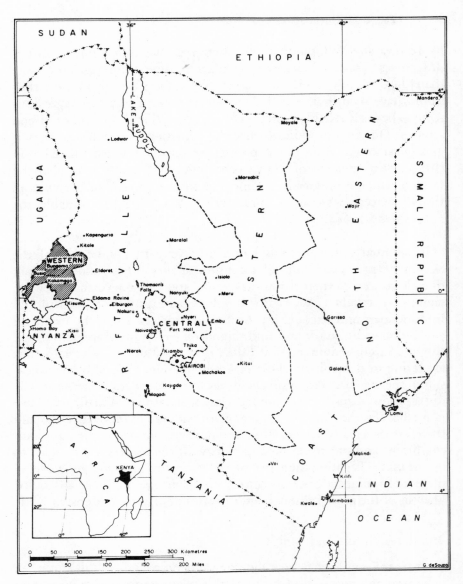

Map 1. Kenya: Provincial boundaries

patterns of staff behavior uncovered in Western Kenya are representative of small-farm agricultural extension not only in the rest of Kenya but also in many other developing countries. As will become evident in the following chapters, the research on agricultural extension in other parts of Kenya, in Tanzania, in Uganda, in Nigeria, and in India provide evidence that the same basic organizational factors are at work.

Western Province is populated almost exclusively by members of the Luhya (Abaluhya) tribe, but it presents a wide range of small-farm conditions. In the southeast in the Vihiga Division of Kakamega District, the population is dense (538 per sq. km. or 1,390 per sq. mile)[19] and the farm sizes are quite small (1.05 hectares or 2.6 acres).[20] Nonetheless, the soil is good and the rainfall heavy (over 1,800 mm. or 71 in. per year),[21] allowing two crop cycles a year. Corn is the staple food crop; bananas are common; coffee and tea are important market crops. About half the farms have one or more cows, overwhelmingly of the local Zebu variety. The cattle are kept basically for their milk, which is in short supply in the area. Much more productive grade milk cows are now being brought into the area and the government is sponsoring an artificial insemination project to upgrade the Zebu stock. Almost all cultivation in the area is done by hand. Because of population pressures and the early availability of mission education in the area, about a third of the male heads of households are away in wage employment.[22]

A clear contrast to Vihiga is provided by the Central and Southern Divisions of Busia District in the southeast of the province on the shores of Lake Victoria. (For the location of these areas and the boundaries of the basic ecological zones, see map 2.) The population is relatively sparse (128 per sq. km. or 332 per sq. mile; farm sizes of 4.5 hectares or 11.4 acres);[23] the soil is less fertile, and the rains are less frequent (about 1,060 mm. or 42 in. per year).[24] Cassava is grown together with corn as a staple food. Cotton is the only significant market crop. Some of the area has trouble with tsetse fly, so keeping cattle is more of a risk and they are less common. Missions did not find these divisions very attractive, consequently the levels of education are relatively low, and there is little male out-migration in search of employment.[25] Despite the large farm sizes, cultivation is done by hand and much of the land is fallow.

Another contrast is provided by Kimilili Division of Bungoma District in the province's northeast. It lies on the borders of the old white settler highlands. Population is not dense (100 per sq. km. or 259 per sq. mile),[26] and farms are relatively large (6.95 hectares or 17.2 acres).[27] The soil is good and the rain is quite adequate for a single

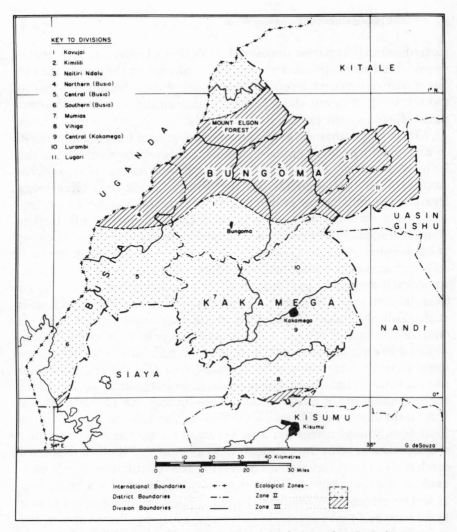

Map 2. Western Province: District and division boundaries and ecological zones

ECOLOGICAL ZONE KEY:

Zone II: Equatorial climate, humid to dry subhumid (moisture index not less than –10). Forests and derived grasslands and bushlands, with or without natural glades. The potential is for forestry (with local wildlife and tourist development) or intensive agriculture, including pyrethrum, coffee and tea at the higher altitudes. The natural grassland, under intensive management for optimum production, supports one stock unit per 1–1 1/2 hectares, dependent on grassland type.

Zone III: Dry Subhumid to semiarid (moisture index –10 to –30). Land not of forest potential, carrying a variable vegetation cover (moist woodland, bushland or "sa-

annual crop cycle (about 1,240 mm. or 49 in. per year).[28] Coffee is grown, but the conditions are ideal for extensive cultivation of corn and the area is a major exporter of corn to other parts of Kenya. Tractor cultivation is common and the division is considered one of the most agriculturally progressive in Kenya.

Western Province is divided into three districts and twelve divisions.[29] We have described four of the divisions, taken from each of the districts. Other divisions would present further variations in agricultural conditions. Nonetheless, enough has been said to indicate that Western Province contains a range of small-farm types almost as great as in Kenya as a whole. In confining ourselves to Western Kenya, we lose little in ecological variety. The variations we have not captured are those in tribal cultures.

The first stage of our research was a pretest of staff and farmer interviews, conducted in North and South Maragoli and West Bunyore Locations in Vihiga Division, Kakamega District.[30] These data were thoroughly analyzed and provided us with many hypotheses that we were to test later.[31]

The major part of our research consists of interviews with a sample of approximately 40 percent of all the junior staff in the province. This sample is not based on a random selection of individuals. As we said at the beginning of the first section of this chapter, we are interested in three units of analysis: (1) the individual extension agents, (2) the small local work groups of agents, and (3) the extension organization as a whole. Although the first and third units could be treated through a random sample of individual agents, the second could not. Adequate analysis of work groups demands data on all the members of the group, as will be evident in chapters 3, 4, and 5. Thus, if groups were to be a unit of analysis we had to make them and not individuals our sampling unit.[32] We knew that the basic junior staff groups are the set of extension agents working in one location, so we used the location as our sampling unit. Twelve locations were selected on a random basis and two others purposefully added.[33] Elgon Location, Bungoma District,

---

vanna''), the trees characteristically broad-leaved (e.g., *Combretum*) and the larger shrubs mostly evergreen. The agricultural potential is high, soil and topography permitting, with emphasis on ley farming. Areas under range-use are still extensive and, under close management, their stock-carrying capacity is high, at less than 2 hectares per stock unit.

SOURCE: Survey of Kenya, *National Atlas of Kenya*, 3d ed. (Nairobi: Survey of Kenya, 1970), p. 28.

was selected because it was ideal for testing a particular hypothesis on staff Innovativeness (see chapter 8). Bukhayo Location, Busia District, was chosen to provide some particular policy-relevant data to the government of Kenya. The data have been carefully examined to determine whether inclusion of these two nonrandomly selected locations biases our results. As they do not, we have included them in our analyses. Our sample includes locations from all of Western Province's divisions except Northern Division, Busia District (which consists of only two locations). (See map 3 for the location of the sampled areas.)

We interviewed all the staff working within the sampled locations and, in addition, all those working on divisional and district duties. Because only a portion of locational staff are included in our sample, against all of the divisional and district staff, the latter are overrepresented. Where possible, we have corrected this by a system of statistical weighting. Comparisons between weighted and unweighted analyses, however, indicate that they produce almost identical results, so where the computer facilities available to us made weighting difficult, we used unweighted analyses. In any case, our basic data are interviews with 169 junior agricultural staff, 44 junior veterinary staff, and 25 senior staff. The junior staff interviews were conducted in the vernacular by three research assistants—Edwin Luchemo, Jack Tumwa, and Humphries W'Opindi. I interviewed the senior staff in English. The following chapters concentrate on the junior agricultural staff, although we do use the senior staff and junior veterinary staff data to elucidate a few major points.

When the basic data had been fully analyzed, we found that a few new and tantalizing questions had been raised to which we had incomplete answers. To complete our picture of junior staff behavior, we conducted a further set of interviews with a portion of our original sample. (The questions asked were new ones. We were not attempting a panel study.) Five locations were selected at random from our original fourteen to form this subsample.[34] Fifty-two interviews with junior agricultural staff were conducted in this final stage of our research in Western Province.

In summary, our research was conducted in Western Province, Kenya, over a two-year period. Data were collected in three distinct stages—the pretest, the basic, and the follow-up phases. The basic data represent a 40 percent sample of the junior agricultural staff in the province, 169 individuals drawn from 14 locations. It is now time to turn to an analysis of the data themselves.

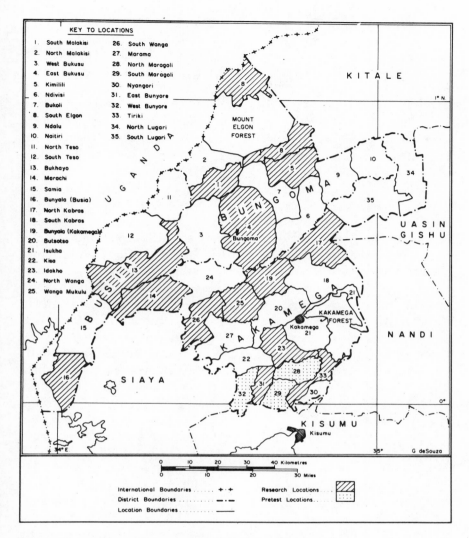

KEY TO LOCATIONS

1.	South Malakisi	26.	South Wanga
2.	North Malakisi	27.	Marama
3	West Bukusu	28.	North Maragoli
4.	East Bukusu	29.	South Maragoli
5	Kimilili	30.	Nyangori
6.	Ndivisi	31.	East Bunyore
7.	Bukoli	32.	West Bunyore
8.	South Elgon	33.	Tiriki
9.	Ndalu	34.	North Lugari
10.	Naitiri	35.	South Lugari
11.	North Teso		
12.	South Teso		
13.	Bukhayo		
14.	Marachi		
15.	Samia		
16.	Bunyala (Busia)		
17.	North Kabras		
18.	South Kabras		
19.	Bunyala (Kakamega)		
20.	Butsotso		
21.	Isukha		
22.	Kisa		
23.	Idakho		
24.	North Wanga		
25.	Wanga Mukulu		

KITALE

MOUNT
ELGON
FOREST

UGANDA

BUSIA

BUNGOMA

Bungoma

KAKAMEGA

Kakamega

KAKAMEGA
FOREST

UASIN
GISHU

NANDI

SIAYA

KISUMU

Kisumu

G. deSouza

1° N.

0°

34° E.

35°

0	10	20	30	40 Kilometres		
0		10		20		30 Miles

International Boundaries . . . . . . + · +   Research Locations . . . .
District Boundaries . . . . — · —   Pretest Locations . . . . .
Location Boundaries . . . . ———

Map 3. Western Province: Location boundaries and research sites

Part 2

Work Groups: Conflict and Leadership

3   Work-Group
   Consciousness

The Social Context of Organization Theory

Why is the performance of agricultural extension agents so poor? Because of inadequate training, low motivation, or bad organization? These are the standard answers and in Part 3 we will demonstrate that they have a considerable impact, although not always in quite the ways that they are usually believed to have. But at least as important a determinant of farmer-contact agent work is the alienation of the agents from the extension organizations in which they serve. Consequently, the social groups of agents are often engaged in collective restriction of their output, and increased supervisory pressure usually results only in a reduction in the amount of work done (as will be shown in chapters 4 and 5). Our objectives in this section are first to examine the group behavior and alienation of junior extension staff and then to see the impact of these factors on the work of the Ministry of Agriculture.

Clearly an organization must be understood before it can be mastered or changed, and no aspect of an organization's dynamics is more important than the structure of its internal cleavages. When we know the patterns of alliances and conflicts, we can anticipate the amounts and sources of opposition and support that the actions of various organizational participants will command. To intervene in an organization without understanding its internal social structure is like setting out on a new journey without map or directions. No experienced administrator doubts the truth of these words when they are applied to the behavior of his peers or to the hierarchical levels above him. Yet the same man all too often is quite unsophisticated about the alliance and conflict-group memberships of his subordinates. The two most common errors are to assume that workers automatically give their loyalties to their superiors or that worker alliances are simply extensions down the hierarchy of conflicts at higher levels. The "one big happy family," the "leader and followers," and even the "factional conflict" models of organizations ignore the fundamental tendency of those subject to a common exercise of authority to band together in self-protection.

The first Western attempt at organization theory ultimately failed because it underestimated the significance of work group solidarity; the subsequent structure of propositions has been built with the group as its foundation stone. Frederick Taylor, the first organization theorist and the founder of the scientific management school of thought, was well aware of the network of social relations among workers. He regarded these ties as essentially undesirable, however, for he perceived that they led employees to be antagonistic to management and to restrict their productivity. Taylor believed that such conflict was unnecessary, arguing, "The principal object of management should be to secure the maximum prosperity for the employer, coupled with the maximum prosperity of each employee" through increases in productivity. "Scientific management . . . has for its very foundation the firm conviction that the true interests of the two are one and the same." Taylor was convinced that he could break the hold of the work group on the individual by tying wage incentives to the performance of well-defined tasks. "It is impossible, through any long period of time, to get workmen to work much harder than the average man around them, *unless* they are assured a large and a permanent increase in their pay" (my italics).[1]

The human relations movement, which succeeded Taylorism, took a fundamentally different attitude toward the work group. This school of thought was prompted by the research of Roethlisberger and Dickson in the Hawthorne plant of the Western Electric Company in Chicago.[2] The Hawthorne studies did not discover the work group, as is sometimes suggested; we have seen that Taylor was well aware of its negative influence. What they did reveal was (1) that the human relations of a work group raises worker morale and thus can increase productivity, (2) that work groups often impose minimum as well as maximum levels of output upon their members, and (3) that workers often will combine to restrict production despite the existence of a well-defined set of tasks and individual wage incentives.[3] Thus Taylor was found to be wrong in two fundamentals: the hold of the group on worker output cannot be broken readily, and destruction of the network of supportive relations among workers is itself likely to be harmful to productivity. The human relations movement led by Elton Mayo has been as persuaded as Taylor of the ultimate harmony of worker and management interests. But as a result of the ongoing chain of studies touched off by the Hawthorne research, Mayo's followers have encouraged management to capture the allegiance of the work group rather than attempt to break its cohesion.[4]

The "informal organization" of workers remains the foundation stone of industrial sociology in the West today.[5]

Worker solidarity implies the existence of an essentially dichotomous pattern of social organization and conflict within the enterprise. Rather than a finely differentiated series of hierarchical gradations and functional specialties, there are two basic, socially significant groups— workers and management. Occasionally the division may be trichotomous, for example, manual labor, clerical staff, and management. Whether the pattern be dichotomous or trichotomous, however, there is a strong tendency in Western work organizations for basic social groupings to form around differences in hierarchical authority and for these groups to be discontinuous, reflecting major social gaps. This pattern of social relationships causes cleavages in the work place to be cumulative rather than crosscutting and leads to the formation of stable conflict groups that are potentially deeply antagonistic. That Elton Mayo's human relations school failed to analyze conflict seriously and instead sought to promote harmony between workers and management should not hide its acceptance of the reality of division between the two and treatment of workers as a potential conflict group whose cooperation requires negotiation.[6] If industrial social groupings were multiple and merged into one another and if they produced crosscutting and fluid cleavages (so that allies on one issue were opponents on the next), it would not be necessary to treat the workers as a group. It would follow that the authority system of the organization would be much stronger and subject to fewer limitations. Thus the usually dichotomous pattern of social organization in Western industry has important consequences for how the enterprise works.

The discoveries of the importance of work-group solidarity have a significant social context. Both Taylor's practical experiments and the Hawthorne studies were done in peak periods of class conflict in the United States. Taylor carried out his series of factory reorganizations at the end of the nineteenth century, when the bitter struggle over the establishment of American trade unions was taking place; the critical study on worker solidarity was done at Hawthorne during the Great Depression, the peak period of socialist party activity in the United States.[7] Of course the importance of work groups has been reaffirmed continuously since that time despite a marked decline in class consciousness in the United States. Nonetheless, the Western countries in which virtually all the research on this subject has been done are characterized by the industrial society occupational structure and have a broadly similar set of cultural values attached to positions in that structure. The question naturally arises whether the dichotomous pattern of work social organization is a cultural or a structural phenomenon. Is worker solidarity largely a function of what can loosely be called the general class character of the Western societies? Or is it instead derived from the basic structural characteristics of a hierarchically organized work situation?

Black Africa differs radically from Europe and America in its class structure. Almost all African political leaders have claimed that their societies are still classless.[8] We need not go that far while agreeing that African class formations are less developed and different in character from those in the industrialized world. In a perceptive essay on Kenya's social structure, Colin Leys concludes, "Whereas the bourgeoisie definitely exists, though with many characteristics that make it rather different from the classical Marxian model, the existence of a proletariat is seriously open to doubt." The workers of Africa are much better seen as urban-based peasants than as proletarians. Furthermore, the elite's extraction of resources from the peasantry has been so impersonal and indirect that the conflict of class interests has not been obvious. Consequently peasant class consciousness and class antagonism have not yet developed.[9] In Kenya

> the character of politics will for some considerable time be determined by the fact that the peasantry as a class has not yet reached the limits of its development, and that the symbiosis between it and the emerging urban-based classes is not yet fully developed either. This symbiosis is mediated by the "dyadic," client-patron relationships which link peasants to the politicians in the towns locality by locality. This corresponds to the continuing importance of the clan or lineage as a unit of social organization; one might say that it is the characteristic mode of political self-expression of peasantries.[10]

Certainly throughout Kenyan society—in national politics, in trade unions, in cooperatives—cleavages are much less likely to be organized around class than around tribes, clans, or ethnic groups.[11]

With the consciousness of class antagonisms weak or absent, we might ask whether the social context exists in Kenya within which the systematic manager-employee conflict of Western theory can become significant. Certainly the agricultural extension service would seem a most unlikely setting for dichotomous conflicts. Its employees are physically based at the very center of peasant society, and the geographical dispersion of their work would seem to give them very little opportunity for collective consciousness or action. If dichotomous tensions are found here, then they are likely to be found in other organizations in Kenya. As Western organizational theory is largely based on the implicit assumption of dichotomous potential conflict groups, their absence in Kenya would throw the applicability of organization theory to that society into serious question. Actually, as we will demonstrate, systematic manager-employee conflicts and work-group solidarity do exist in the extension services and in Kenyan organizations in general. The existence of such conflict groups and their particular character have great significance for the way these organizations work.

Conflict-Group Theory

Before we can analyze the social structure of the agricultural extension services, we require a theoretical framework. This is not a solely academic requirement. Since the character of an organization's conflict-group structure has important implications for the way it functions, we need precise guidelines concerning what to look for as we try to classify it. For a variety of reasons, the best theoretical framework for analyzing organizational conflict groups is the one provided by Ralf Dahrendorf in *Class and Class Conflict in Industrial Society*. Most importantly, he focuses on precisely the issues of the definition of, consciousness of, and conflict between subordinate and superior work groups which we believe are so crucial. Another advantage for our purposes is that Dahrendorf defines industrial conflict groups according to the possession of authority rather than the ownership of property. This makes it possible to apply a class type of analysis to government organizations, where the concept of ownership is inappropriate.

A final point in favor of Dahrendorf's theory is that it is based on Karl Marx's classic analysis of class and demonstrates the analytic similarity between conflict analysis at the societal and organizational levels. Although there is no necessary link between worker-management conflicts within organizations and class conflict in society as a whole, the character and intensity of the two do seem related. For example, the development of strong trade unions represents a similar stage of conflict in both arenas. Furthermore, a work organization in many respects is a microcosm of society. Virtually all the questions about the character of conflict groups that class theorists from Marx onwards have asked at the societal level are worth asking in the context of a single organization.

We do not believe that Dahrendorf's adaptations of class theory are appropriate at the macrolevel, but this is a question that does not concern us here; we need only accept his formulations as a fruitful set of analogies for intraorganizational study. In the following analysis, we will unabashedly apply at the organizational level statements originally made about societies in order to exploit these analogies.

Dahrendorf focuses on "social conflict groups the determinant . . . of which can be found in participation in or exclusion from the exercise of authority. . . ." [12] His argument is that "in every social organization some positions are entrusted with a right to exercise control over other positions" and that "this differential distribution of authority invariably becomes the determining factor of systematic social conflicts." [13]

Dahrendorf works with a two-group model of social conflict. There are those who occupy roles of authority, or superordination, and those who lack authority and are subordinates. He is uninterested in conflicts between those who hold authority but in differing degrees (for example, between the middle and upper levels of a hierarchy). [14] To hold rigidly to such a position seems unnecessary. It can be argued instead that although conflicts based on authority may be trichotomous as well as dichotomous, both will differ fundamentally from situations in which there are many continuous groups with crosscutting cleavages and fluid patterns of conflict. [15] In this particular study we actually use a two-group model. The foci of our interest are those at the bottom of the hierarchy and their relations with those who exercise authority over them. Dichotomous types of conflict probably occur further up the hierarchy, however, and would have to be analyzed in a detailed study of senior staff behavior.

It is one thing to say that systematic conflict can be based on differences in the distribution of authority and another to say that they are. Dahrendorf distinguishes between latent and manifest interests and between quasi groups and interest groups. People who occupy a common position toward authority have a set of *potential* mutual interests. If they are not conscious of these interests or of themselves as a group, their mutual interests are only latent and they are a potential group, a quasi group. When a common consciousness is generated, manifest interests and an interest group develop. This group consciousness is the precondition for dichotomous conflict. [16]

The existence and character of group consciousness determine whether we should apply stratification or class theory to our analysis of a given situation. Virtually all social settings reveal a finely graded hierarchy of differences among people, whether based on income, status, authority, or whatever. Such a hierarchy becomes a dichotomous conflict system only when there is a clear break (or perhaps breaks) in it so that several gradations on each side come to be grouped together and available for common struggle. Where instead we find an uninterrupted continuum of social differences, the implication of social mobility, a shared set of values throughout the hierarchy, and unstable, less systematic conflict, stratification theory provides a more accurate fit.[17] We can illustrate the difference between class and stratification approaches by a military example. An army is made up of a large number of finely graded hierarchical ranks. The differences between lieutenants, captains, and majors are ones of stratification. Mobility is relatively easy from one rank to the next, and whatever their conflicts, these ranks basically identify with one another. Below the lieutenant comes a major break, however, based upon the difference between commissioned and noncommissioned ranks. The hierarchical distance between a lieutenant and a captain, on one hand, and between a sergeant and a lieutenant, on the other, are the same. But the social gap is much greater where rank is compounded with the status of commissioned officer. Mobility within the two categories of commissioned and noncommissioned is frequent and routine, but movement from one to the other is rare and difficult. Recruitment to the two categories differs in systematic and important ways, compounding the potential for conflict across this particular hierarchical gap. Within either the "officers" or "men" group, stratification analysis can be applied; between them, dichotomous group analysis and the study of systematic social conflict is required.[18]

To have full consciousness, a group must have both a culture and an organization for conflict. In this way it moves from being a quasi group to an interest group. In *The 18th Brumaire of Louis Bonaparte*, Marx writes,

> In so far as millions of families live under economic conditions of existence that separate their mode of life, their interests, and their culture from those of other classes and put them in hostile opposition to the latter, they form a class. In so far as . . . the identity of their interests begets no community, no national bond, and no political organization among them, they do not form a class.[19]

Regarding interest-group culture, Dahrendorf quotes approvingly from an early essay by J. A. Schumpeter, "the members of a class behave towards each other in a way characteristically different from that towards members of other classes, that they stand in a closer relation to each other...."[20] To this we would add that members of an interest group that is concious of itself will treat people of other strata *within* their interest group quite differently from those of other, similarly distant strata *outside* their groups.

For an interest-group culture to develop, organizational recruitment and allocation of roles must be socially stable. Dahrendorf stresses these two features by discussing their absence. If mobility is such that the placement of people at different levels of the organizational hierarchy bears no relationship to their social origins, then an interest-group culture is unlikely to result.

> Empirically, the formation of organized interest groups is possible only if recruitment to quasi-groups follows a structural pattern rather than chance.... Persons who attain positions relevant for conflict analysis not by the normal process of the allocation of social positions in a social structure, but by peculiar, structurally random personal circumstances, appear generally unsuited for the organization of conflict groups.[21]

The latent interests created by differences in authority are unlikely to become manifest unless they are compounded by consistent differences in social origins or career prospects between the two groups. The second point concerns the organization's division of labor, for if this is unstable, members will be unable to construct permanent alliances. "If imperatively coordinated associations are either themselves just emerging or subject to radical change, the probability is small that the quasi-groups derived from their authority structure will lead to coherent forms of organization."[22]

The final step in the transition from quasi group to interest group occurs when a group with consciousness becomes organized as well. This last step does not always occur. Marx stresses this point particularly for peasants, whose isolation makes difficult the social interaction necessary for organization.[23] When this final transition does take place, however, some form of conflict of the dichotomous type is inevitable.[24]

The Existence of Interest Groups in the Ministry

Does the potential for dichotomous conflict exist in the Ministry of Agriculture in Western Province? We have noted that this is an important question and that, because of the general weakness of class consciousness in Kenya, we cannot assume an affirmative answer. Drawing on Dahrendorf, we have provided ourselves with a definition of interest group and the means for analyzing the extent of interest group formation in an organization. We are now prepared to demonstrate that the answer to the question is Yes.

The analytic decision about who belongs to the quasi group of "dominators" and who to the "subjugated" is fairly easily made. Those whom the ministry terms its senior staff have authority; those called junior staff do not. The senior staff consists of Agricultural Officers (AOs), Veterinary Officers (VOs), Assistant Agricultural Officers (AAOs), and Livestock Officers (LOs) (in short, those whose title includes the word *officer*). As far as we know, a man or woman in this group always has someone under his or her direction (see figure 1). The junior staff are made up of Agricultural Assistants (AAs), Animal Health Assistants (AHAs), Junior Agricultural Assistants (JAAs) and Junior Animal Health Assistants (JAHAs). We have never encountered a JAA or JAHA in a supervisory position; they are at the bottom of the organizational hierarchy. Most AAs and AHAs also are without any claim to authority.

A few AAs and AHAs are first-line supervisors, however. Most commonly this is the LAA, who is the AA in charge of the small team of AAs and JAAs assigned to a particular location. In addition, a few AAs and AHAs head three or four junior staff at work on a specialist task (such as coffee extension or veterinary work). Do these junior staff supervisors belong to the quasi group of dominators or subjugated? This is difficult to answer in the abstract. First-line supervisors are a border category, sometimes identified with the management, other times with the workers.[25] Dahrendorf would probably classify them with the senior staff, for he considers a position with any authority to be one of domination.[26] We think it sensible to be more flexible on this point. Looking at the absence of authority as relative, we feel it appropriate to use additional objective factors, such as recruitment group and terms of service, to define the exact boundaries between quasi groups. In this particular case, the first-line supervisors seem to us to be members of the quasi group of junior staff. The LAA, for example, is drawn from the general group of AAs, can be returned to it with relative ease, and receives no additional pay for his responsibilities. Furthermore, the amount of authority he exercises over his colleagues is quite limited: he continues to do the same work that they do, communicates to them the

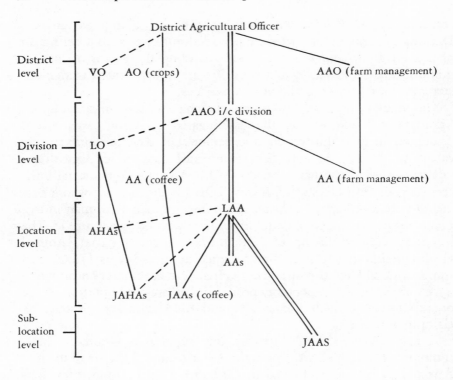

Key:
_____ a line of formal and actual authority and responsibility
======== the line of basic, unspecialized authority
-------- a line of formally established but practically weak authority and responsibility
AO   Agricultural Officer
VO   Veterinary Officer
AAO   Assistant Agricultural Officer
LO   Livestock Officer
LAA   Agricultural Assistant in charge of a location
AHA   Animal Health Assistant
AA   Agricultural Assistant
JAHA   Junior Animal Health Assistant
JAA   Junior Agricultural Assistant

Fig. 1. A typical organization chart for the Ministry of Agriculture in Western Province

instructions of the AAO, and reports verbally on their work performance. He is unable to apply or recommend sanctions himself.
Compared to that held by senior staff, his authority seems minor and so
we include the first-line supervisors in the junior staff quasi group.

Junior and senior staff in the ministry differ markedly in their terms of service. The general rule is that junior staff are assigned to their home area, or as close to it as possible; senior staff may be assigned to an area where they understand the vernacular but almost never to their exact place of origin. Senior staff are entitled to government housing and official Land Rovers; junior staff rarely receive housing and virtually always provide their own transport, usually a bicycle. There is also a distinct break in the range of salaries between the two groups. This is not immediately obvious from an inspection of the current salary structure of the ministry (table 5). Each cadre of employees (JAAs, AAs, AAOs, AOs) begins service at a bit more than half the salary of the cadre above it. The differences between the cadres are treated as equal steps on a hierarchical ladder. But regarding qualifications, the steps are not equal. The differences in years of general and technical education are two between AOs and AAOs, one between AAOs and AAs, and four or more between AAs and JAAs. The gap in qualifications is markedly the smallest between AAOs and AAs, at the boundary between senior and junior staff. But it is also here that the proportional gap in salaries is marginally the greatest. This equal treatment of differences that are not equal highlights the status gap between the two quasi groups of junior and senior staff. This gap has its origins in the colonial system, where junior staff were invariably black and senior staff white.[27]

Having identified a distinct break in the economic and work conditions of the two quasi groups, we proceed to look for the systematic differences in social interaction that would indicate a group culture. This can be studied through the friendship choices of both staff groups. At the end of each of our interviews, which were very much work oriented, we asked, "Now finally, we find that an extension agent's work is often helped or hindered by his personal relations with those around him. For this reason we would be grateful if you would name for us your friends whom you see regularly." Where the respondent was unclear, we stated that we were interested in his friends in the general geographical area and that our question included all types of friends. After this first query was answered, we probed with "Now, in addition, what (other) friends do you have in the Ministry of Agriculture?" In total we recorded up to fifteen friendship choices, of which no more than ten were from outside the Ministry of Agriculture. In only a very few cases were these upper limits reached. In addition, we ascertained the nature of each friend's occupation and where he lived. Cooperation in answering all these questions was generally very good. In using this sociometric data to describe the informal social system of the Ministry of

TABLE 5
The Current Salary Scales of the Ministry of Agriculture

Job Title	Job Qualifications	Years of General and Technical Education	Range of Salary	Starting Salary as a Proportion of That of Next Highest Cadre	Maximum Salary as a Proportion of That of Next Highest Cadre
Junior Agricultural Assistant	Certificate of Primary Education or Junior Secondary Certificate	7 or 9	K£ 201– £417 (U.S. $575–$1,190)	.54	.60
Agricultural Assistant	Secondary School Certificate and a 2-year Certificate in Agriculture	13	K£ 369– £690 (U.S. $1,055–$1,970)	.52	.55
Assistant Agricultural Officer	Secondary School Certificate and a 3-year Diploma in Agriculture	14	K£ 714– £1,254 (U.S. $2,040–$3,590)	.59	.54
Agricultural Officer	University Degree in Agriculture	16	K£ 1,212– £2,334 (U.S. $3,465–$6,670)		

Agriculture, we do not want to imply that social structure consists only of friendship patterns. This is obviously not the case. When someone claims another as his friend, he is not saying either that he sees that person often or that he does not interact frequently with others. A friendship choice indicates only that he would like to have contact with the person, whether he actually does or not. Nonetheless, this information is extremely useful in locating the boundaries of people's affections, which in turn is helpful in identifying status and other barriers between people.

The responses to the friendships question immediately reveal major differences between the styles of life of junior and senior staff.[28] The latter are enmeshed in an almost exclusively civil servant social circuit, where the former are involved in the social life of their rural communities (see table 6.) The relative isolation of the senior staff probably results largely from their being away from their places of origin and living in government staff compounds.[29] The customary senior staff club at district headquarters also provides a center for social activity that is usually unavailable to junior staff.

TABLE 6
Work Places of Friends Named by Junior and Senior Agricultural Staff

Staff Group of Respondent	Work Place of Friend			
	Ministry of Agriculture	Other Government	Outside Government	Total
Senior staff (N = 25)	51%	37%	12%	100%
Junior staff (N = 169)	24%	35%	41%	100%

The social life of junior staff is focused on their local areas. Consider the friends they name who do not work for the Ministry of Agriculture. On average, 45 percent live in the extension agent's home sublocation, and 30 percent more are from within his location. That 25 percent of his friends reside beyond the boundaries of his location marks the AA or JAA as both more cosmopolitan than his farmer neighbor and much more parochial than his senior staff superiors.[30]

What is the social status of the friends with whom the junior staff interact socially, and, by inference, what social status do they assign to themselves? We asked respondents to tell us what kind of work each friend does. On this basis each nonministry friend was assigned to one of four predetermined status categories and the percentage of friends in these categories was calculated for each respondent.[31] Table 7 defines the four categories and gives the average percentage of friends in each one. From these figures it seems clear that junior staff see themselves as part of the rural elite, but in the lower or middle part of that group. The data confirm our impression that they belong to a status a bit lower than that of a primary school teacher. The agricultural extension agent is probably often the social equal of the province's progressive farmers, who fit into the Lower Elite status group.[32] These data support the conclusion of Thoden van Velzen for Rungwe, Tanzania, that staff associate very largely with the richer peasants in their social contacts with farmers.[33] The approximately 90 percent of the rural population that falls into the Nonelite status receive only 20 percent of the friendship choices.

TABLE 7
The Average Percentage of Nonministry Friends Chosen from Various Status Categories by Junior Staff

Percentage	Category	Exemplary Definition
7	Upper Elite	Chiefs, headmasters, County Councillors, big businessmen, other relatively well-to-do
39	Middle Elite	School teachers, subchiefs, moderate businessmen, big farmers, middle salaried group
33	Lower Elite	Small businessmen, traders, moderate farmers, lesser employed
20	Nonelite	Average farmers

Where the junior staff belong to the rural elite, the senior staff are members of a national elite, what John Okumu labels the salariat. This new African middle class has inherited the status, power, and much of the wealth of its European colonial predecessors. As the label indicates, its present economic strength lies in its high salaries and not in ownership of productive property.[34] The immersion of senior staff in a salariat social system can be seen in the proportion of their friends—88 percent— who work for government (table 6) and in the social status of their friends (table 8). Sixty-six percent of the friendship choices made by senior staff are of people equivalent in status to themselves, the highest status group in the area. They have only 19 percent of their friends among the lesser employees, traders, and farmers. These occupations belong to the Lower-Elite and Nonelite status groups, in which the junior staff find 53 percent of their friends (table 7). It is true that social life for the divisional level senior staff is less exalted than for those at district headquarters, but they still draw only 30 percent of their friends from the Lower Elite and Nonelite status groups.

TABLE 8
The Average Percentage of Nonministry Friends Chosen
from Each Status Category by Senior Staff

Status Equivalent of Friends	Respondent		
	District Senior Staff (N = 13)	Divisional Senior Staff (N = 12)	All Senior Staff (N = 25)
District head of department	61%	13%	37%
Divisional head of department or district aide	22	36	29
Chiefs, teachers	10	20	15
Lesser employees, traders, farmers	7	30	19
	100%	100%[a]	100%

[a]The apparent total of 99% derives from rounding errors.

The foregoing material does indicate that the quasi groups of junior and senior staff "live under economic conditions of existence which separate their mode of life, their interests, and their culture" from one another, to use the words of Marx.[35] But do the members of each quasi group, as suggested by Schumpeter, "behave towards one another in a way characteristically different from that towards members of other" quasi groups?[36] This question can be answered by analyzing the choices of friends within the Ministry of Agriculture. Tables 9 and 10 present data on the friendship choices made and received by the various hierarchical and status levels—senior staff, supervisory AAs, AAs, and JAAs.[37] Interpretation of these tables requires that we keep in mind the usual social "tendency for lower-status group members to direct their friendship choices disproportionately to upper-status members," and that "upper-status members tend not to reciprocate but to direct their choices to others also high in status."[38] Thus the naming of AAOs as their friends by some AAs and JAAs would not prove the absence of a dichotomous barrier. Neither would a failure of most senior staff to choose their juniors as friends prove that a barrier existed. Instead, within the expected general pattern of juniors looking upward in the hierarchy for friends and seniors not reciprocating, we are seeking a distinct change in the patterns of response to status and hierarchical position as the quasi group boundary is crossed. In fact, we do find such

TABLE 9
The Frequency with Which One Is Named a Friend by Junior Staff

Type of Friend Named	Average Times Chosen	Number in Category
Non-Luhya senior staff	1.79	14
Luhya senior staff	4.48	11
Supervisory AAs	4.21	34
Other AAs	3.82	45
JAAs	2.16	89

TABLE 10
The Frequency with Which One Names Those at Lower Levels as Friends

Respondent	Average Number of Choices	Number in Category
Non-Luhya senior staff	.36	14
Luhya senior staff	1.45	11
Supervisory AAs	2.41	31

a break in these tables. Table 9 shows that where popularity with junior staff increases with status inside that group (reaching an average of 4.2 choices received by supervisory AAs), it drops off as one passes over to the senior staff (who receive an average 3.0 junior staff choices). From table 10 we also learn that supervisory AAs name AAs and JAAs as friends almost four times as often as do the senior staff (2.41 vs. 0.68). Junior and senior staff groups each seem to be conscious of their own identities and their distinctness from one another.

Nonetheless, there is an important identity that crosscuts that of class in Kenya. If we were to stop our analysis at this point, we would be simply assuming that tribe has no significance in organizational relationships. In fact, when we examine the impact of tribe and quasi group together as sources of division, our picture of the ministry's social system changes considerably. Because of the different policies on job postings mentioned earlier, virtually all of the ministry's junior staff in an area belong to the local tribe (in this case the Luhya), while a majority of the senior staff will be outsiders. In Western Province, 44 percent of the officers at district and divisional level are Luhyas. Looking again at table 9, we see that the mass of AAs and JAAs see the Luhya senior staff as simply the highest status level within their own social group. They feel the class break strongly only with the non-Luhya senior staff.

The same pattern is shown in table 11, where staff friendships with the members' own immediate superiors are shown. Within the two quasi groups, the proportion of subordinates naming their supervisor as a friend ranges between .32 and .36. Between the two quasi groups, the proportion drops to .22. But when the superior is of the same tribe as his staff, 28 percent name across the quasi group boundary. When tribe and quasi group barriers are combined, only 17 percent make a friendship choice.[39] Differences in quasi group and tribe together create social obstacles for the junior staff, but quasi group by itself does not seem to have much effect. Quasi group has much more significance for the senior staff. Where supervisory AAs name an average of 2.41 AAs and JAAs as their friends, Luhya senior staff choose an average of only 1.45 JAAs, AAs, and supervisory AAs, and non-Luhya officers select .36 (table 10). Thus, when we are looking downward in the organizational system, quasi group and tribe have approximately equal significance as social barriers.[40]

TABLE 11
The Proportions of Subordinates Who Name Their Immediate Superior as Their Friend

Superior	Immediate Subordinates	Proportion of Subordinates Naming Superior as Friend	Number of Subordinates
LAAs	AAs + JAAs	.32	134
Luhya AAOs	Supervisory AAs	.28	15
Non-Luhya AAOs	Supervisory AAs	.17	12
DAOs[a]	AOs + AAOs	.36	22
PDA[b]	DAOs	.35	3

[a]District Agricultural Officer.
[b]Provincial Director of Agriculture.

The social behavior of the junior and senior staff in Western Province indicates that they are conscious of their group differences. The system is crosscut by tribal identities, however, and this weakens interest-group consciousness, although more for the junior than the senior staff group. A certain amount of clear interest-group consciousness exists, mainly because the junior staff belong to one tribal group while most of the province's senior staff and, of course, the Nairobi leadership do not. This sort of interest-group system probably exists in most of Kenya's rural-based ministries. There is a distinct break in cadres posted nationally and locally in, for example, Provincial Administration, Education, and Cooperatives. Only when both authority and non-authority roles are held by a good mixture of tribes would it be even possible for the interest-group system to be sufficiently crosscut by tribal alliances to become inoperative.[41] In chapter 11 we analyze a number of other Kenyan case studies bearing directly on this point. Our conclusion is that superiors *can* create cleavages on ethnic lines that will override the natural tendencies toward authority conflict in their organizations. Nonetheless, authority conflict is usually dominant because its repression requires extremes of tribal patronage by superiors that they are only seldom willing and able to exercise. Thus one is fairly safe in assuming that authority is the most important source of intraorganizational cleavage, in Kenya and presumably elsewhere as well.

Earlier, we noted Dahrendorf's conclusion, "formation of organized interest groups is possible only if recruitment to quasi-groups follows a structural pattern rather than chance."[42] By this he means that conflicts along authority lines are likely to be significant only when those in superordinate and subordinate positions in an organization are customarily each recruited from different sections of society. Recruitment to authority positions in Kenyan society depends overwhelmingly on examination achievement in the formal education system. At the moment, one examination largely determines who will enter each of the ministry's cadres—the Secondary School Certificate examination, which is taken at the end of eleven years of education. An excellent performance in the sciences will enable one to enter higher secondary school, with the strong probability of going on to university and becoming an Agricultural Officer. A good performance admits one to Egerton College for a diploma and the rank of an AAO. A moderate pass qualifies one for AA training at the Embu Institute of Agriculture. Failure or a poor pass qualifies one for nothing, and one might enter the ministry as an untrained JAA. Achievement on this examination is largely related to the personal capacities of the individual, although there is a good deal of chance and imprecision in any such measurement. It is also related to the social structure of the rural areas in a way that is becoming increasingly clear.

At the end of the colonial period the relationship of educational achievement to social origins was unstable.[43] Because of racial barriers in the European dominated wage sector, the best opportunities for wealth for Africans were in private business and farming, where school accomplishments are not important. Now that employment has again become the preferred avenue to success,[44] educational achievement is again clearly related to the social position of parents. This point is vigorously denied by the Kenyan elites, who see themselves as a meritocracy and point to their origins in small-farm families.[45] It is becoming increasingly clear, however, that these small-farmer parents generally had significant advantages over their neighbors in wealth and status and were often their employers as well.[46] Thus the appearance of structural randomness in recruitment to the senior civil service in Kenya is deceptive. As is true in most countries, positions are related to social class of origin. Within organizations such as Agriculture the dissimilarities in social origins are reinforced by the different pattern of postings for the two quasi groups, which causes tribal differences at the local-level authority boundary. Thus, for the ministry in Western Province, as in most organizations, class rather than stratification theory provides a better framework for conflict analysis.

Interest-Group Cohesion

Interest-group consciousness does not lead to organization and conflict of itself. Located in the rural areas as the extension services are, physical distances create barriers to social interaction. Marx attributed the lack of peasant class organization in France to a similar difficulty.[47] Junior staff are widely dispersed in their places of working and living, but unlike most peasants, they do come together for staff meetings, pay days, training courses, and the like. Thus, an average of 24 percent of the friends of junior staff work for the ministry. This social involvement in the ministry is weak compared with that of workers in a factory setting, but it is still psychologically significant. No fewer than 85 percent of the junior staff are named as a friend by at least one other extension agent.[48] Forty-six percent of a man's ministry friends work in the same location as he does, and this work unit seems to be the focus of his allegiance. Nonetheless, he has a number of ties with others in the same division (an additional 27 percent) and even with those in other parts of his district and the province (the remaining 27 percent). These latter friendships are usually formed on staff training courses at the Bukura Farmer Training Centre. Ministry friendships outside the province are very rare.

The location work group averages a manageable ten members (including the LAA) but achieves real social cohesion only rarely. Kerlinger suggests that we measure the cohesiveness of a group by the proportion of reciprocal friendship choices made out of the number possible.[49] The average on this measure of the fourteen location groups we sampled is .06 (with a range from .00 to .25). Another way to measure the same phenomenon is to give the average proportion of other group members whom individuals name as their friends. Here the average is .26, with a range of .06 to .46.[50]

The senior staff are decidedly more cohesive than their juniors, which is, as Mosca notes, a normal advantage of those at the top of the hierarchy.[51] The proportions of reciprocal friendship choices among the headquarters' staff of the three districts are .30, .33, and .17. The figures for the whole senior staff in these districts are .19, .13, and .06. On the other indicator of cohesiveness—the average proportion of other group members named as friends—the headquarters' measures are .42, .50, and .30, and those for the full districts are .36, .30, and .16. The involvement of the divisional AAOs and LOs in a district-wide senior staff social system is clear, although it is weak. This is not surprising, since senior staff would need transportation to reach their counterparts, and gasoline is a notoriously scarce resource in the ministry. Although the cohesiveness for two of the three headquarters teams is moderately good, it does seem low for groups that share common officers, a common speciality, and common problems. The high rates of transfer in the Kenyan senior civil service doubtless depress the levels of group cohesion.

The seasoned Kenya observer would probably also suggest that tribal conflicts among the officers are reducing their unity. Contrary to such expectations, we did not find tribe having a systematic effect on senior agricultural staff relationships in the province. There were certain individuals who were liked or disliked by their fellows partly because of their tribe, but among the 25 officers whose friendship choices we analyzed, we found no overall tribal pattern in their personal relations inside the ministry. It is only when we examine friendships with those in other ministries and outside of government that we find the pull of tribal affiliation. If we look down each of the columns giving the "Tribe of Friends" named in table 12, we consistently find that those with the largest proportion of friends of a given tribe belong to that tribe themselves. (The only exception is the "Others" column, made up of several tribes and therefore without meaning in this regard.) Statistical analysis of the table indicates that all the tribal groups of senior staff are biased toward their tribemates to at least some degree in their choice of friends. On average, the probability that a senior staff member will discriminate in favor of his fellow tribesmen in his choices is + .19.[52] This tendency is not strong but it is significant. Of course this is precisely what we would expect—people will have a natural pull toward those who have the same language and customs, can share rides home for holidays, and sometimes have news of friends and relatives. In fact, the tribal pattern of friendships evidenced here is much less than the ever-present talk of tribalism in Kenya would suggest. Nonetheless, the existence of ready-made tribal friends in other ministries and outside probably provides social outlets that relieve the urge of work mates to visit with one another and so reduces what might otherwise be a strong senior staff cohesion.

TABLE 12
The Average Percentage of Nonministry Friends in
Each Tribal Group for Each Tribal Group of Senior Staff

| Tribe of Respondent | (N) | Tribe of Friends | | | | Total |
		Luhya	Kikuyu	Luo	Others	
Luhya	(11)	61	8	12	19	100%
Kikuyu	(6)	30	36	11	23	100
Luo	(4)	34	31	26	9	100
Others	(4)	55	13	13	19	100
All	(25)	48	17	16	18	100%[a]

[a]The apparent total of 99% derives from rounding errors.

Summary

Interest groups have developed within the Ministry of Agriculture in Western Kenya. This interest-group formation has been possible partly because the senior staff are tribally diverse while the junior staff are not. The existence of tribal differences along the quasi group boundary gives the junior staff a stronger sense of subordinate identity than they would have otherwise. But, although the junior staff work groups are real social units, they are generally not very cohesive, because the extension agents are geographically dispersed. The personal ties that would be necessary for action on a district, provincial, or national basis are very few. The senior staff are more cohesive than their subordinates, although the level is still only moderate. They also have a much stronger consciousness of their quasi group identity. Their better unity should give them a decided advantage in any dichotomous-type conflicts with the junior staff. It is to the questions of dichotomous conflict and collective action by the junior staff that we turn in the next chapter.

4 Anonymous Group Conflict

Collective Resistance

The ultimate importance of the existence of superordinate and subordinate interest groups based on differences in authority lies not in their existence but in their consequences for the way the organization operates. In chapter 2 we demonstrated that the level of performance of agricultural extension agents is disturbingly low. For example, agents average only 1.75 farm visits a day, although there is broad consensus that 4 to 5 is a reasonable number. The existence of subordinate interest groups, the way in which they operate, and the manner in which they are handled are directly related to this poor level of Visit Effort. These are the issues which we now need to analyze.

At the level of overt behavior, the subordinate interest group of junior agricultural staff is not available for collective conflict with the superordinate interest group of senior staff. First, the great majority of agricultural extension agents are neither members of nor interested in the Kenya Civil Servants Union, which is supposed to cater to their interests. Nonetheless, it is always possible in Kenya that worker rejection of a union is due to inadequacies in its structure or leadership rather than to a disinterest in collective action.

Vigorous trade union activity is the archtypical manifestation of subordinate interest-group action and of dichotomous conflict within an organization. Yet it is not the only form. A formal strike has a number of analytically separable elements—conflict between the management and workers, open and vigorous prosecution of the conflict, collective as opposed to individual action on the part of the workers, capacity of the subordinates to exercise collective control over themselves, willingness of the workers to support one another against individual management reprisals, and formal organization of the subordinates. Other forms of interest group action combine some or all of these same elements. In the case of the Ministry of Agriculture, there have been no extension agent strikes, there is no effective formal organization of subordinates, and whatever conflict exists is not often open.

Granting these points, there are a number of important questions still to be answered about the interest group of junior staff and their conflict behavior. Do they have any *informal organization*? Are they willing to protect one another from senior staff discipline, providing them with a kind of *collective autonomy*? Can they exercise *collective control* over themselves? Are there acts of *individual resistance*, which would indicate an underlying awareness of a conflict of interest with senior staff? Is there any *collective resistance*? Answers to these subsidiary questions will provide a picture of the patterns of dichotomous conflict in the ministry and their importance for organizational behavior.

To gain information about collective and individual resistance, we interviewed the fifty-two junior staff working in five locations selected at random from our original sample. These interviews furnish definite evidence of collective resistance, but they also reveal a strong desire on the part of junior staff to hide their collective actions from public view. The question which brought the most revealing answers in these two regards is: "In your experience of working with the Ministry of Agriculture have you ever seen a situation when extension workers in some way agreed to restrict their work against the wishes of their superior?" Eleven separate incidents are identified by the junior staff, dated from 1953 to 1972. Six occurred in the colonial period under European supervisors; five since Independence under African ones. Eight of these confrontations are reported to have involved informal organization on the workers' part, forcing the defeat of the superiors in six of them. (The junior staff were thwarted in the three cases where they had no organized response.) The incidents were usually over fairly minor issues. In 1972 the junior staff in one area refused to contribute for a party in honor of an outgoing AAO. In 1969 extension workers in training at an FTC would not do what they saw as an excessive amount of work and were finally backed up by the Provincial Director of Agriculture. In 1958 agents resisted an order to carry *jembes* (hoes) with them on their rounds and were finally supported by the Director of Agriculture. Where strong worker resistance was broken, it was usually done by promoting those who complied and transferring or sacking the resisters. Such a history certainly seems to indicate some potential for interest group action by the junior staff. The significant thing, however, is that not one of these conflicts is reported by more than two respondents, and only twelve out of the fifty-two interviewees mention any incidents at all. Are our informants offering us fictional accounts then? It seems possible that one or two of the incidents occurred largely in the respondent's imagination, but most of them look genuine. It appears that many of those interviewed do not report conflicts they personally experience. After all, even those incidents confirmed by a second informant are not mentioned by the others in the same locational work group who must have experienced them.

The desire of junior staff to deny any personal involvement in collective resistance, even when they actually would be willing to participate, is brought out in the responses to another pair of questions. We presented our interviewees with a story in which an AAO announces that he is requiring a doubling of work from his extension agents (which is actually not an unreasonable demand, as we shall see later). The AAO also introduces a realistic system for checking on the work done, another innovation. After the meeting the junior staff talk about the new demands among themselves. A young AA proposes that the agents band together and continue to do exactly equal amounts of work at the old level, while an old AA argues that opposition is immoral and that they should comply. The respondents were then asked whether they personally thought the young or the old AA was right. The responses are overwhelmingly in favor of compliance and against collective resistance (49 out of 52). Yet, when asked to predict the behavior of their fellow agents, fully 50 percent of the junior staff indicate that they would expect some form of open or covert resistance. It is clear that the agents do not want to acknowledge publicly their actual personal willingness to engage in collective action.

Thus there is interest-group conflict among the ministry's field staff, but it follows an atypical pattern. Many of the junior staff are willing to engage in collective resistance against their superiors and to organize reasonably effectively for this purpose. Their strategy, however, is to dispute the existence of such conflicts, often even while they are taking place. Workers may be willing to admit to their supervisors that their fellows have grievances that need correction but they are likely to deny completely that any collective resistance is occurring. These semistrike actions take on an anonymous character. Rarely will anyone admit that they are occurring; discovery of their leadership is almost impossible (as we shall see later in this chapter). The greater part of this anonymity is derived directly from junior staff fear of the tremendous sanctions at the disposal of their superiors—dismissal, unpleasant transfers, refusal of promotions, denial of salary increments. Except for dismissal, these sanctions can be applied fairly easily. The AAs and JAAs are being paid above-average wages for their educational level. Given the large amount of unemployment in Kenya and the specialized nature of their skills, they would find it virtually impossible to find a comparable job if they were fired. Thus even a remote threat of dismissal is greatly feared. The sanction which can be applied almost at whim—transfer—also carries fairly severe economic penalties. When an agent is assigned to his home sublocation, he can live in his own house and supervise his farm. If he is transferred far enough away, he will have to leave his family, rent accommodation, and hire more labor for his farm. This will probably cost him an additional 60 shillings per month, or about 15 percent of the salary of a JAA. Not only are the sanctions that threaten extension agents major ones, but African senior staff have often followed their colonial European predecessors in wishing to crush completely any who show overt signs of resistance to their authority. Covert resistance both protects its organizers and avoids triggering the vindictive anger that superiors in Kenya sometimes show when they feel that their authority is threatened. The meekness of the Kenyan worker before his supervisor may be genuine for many but probably for the majority it reflects more strategy than substance and masks the collective resistance beneath. Such covert resistance is actually much more difficult to deal with. The general lesson to be learned is that the apparent absence of interest-group conflict in an organization is more likely to mean that it has been driven underground than that it is missing.

Group Control of Output

Even better evidence of the potential of the junior staff for interest-group action is found when we move from the issue of direct confrontation (open or covert) with the orders of superiors to the question of group control of output. The classic way that a work group demonstrates its collective autonomy from managerial control is by developing its own standards of a "fair day's work" and bringing pressure on its own members to keep to them.[1] We have found clear evidence that many of the locational groups of agricultural extension agents do exercise control over the amounts of work done by their members.

American research indicates that group controls on the rate of production typically involve limits on the amount of effort workers put into their tasks but not on the amount of skill they have in their jobs.

Restrictive work groups put pressure on those who do "too much" work but are apparently still able to appreciate those who are highly capable.[2] Several indicators of the output of agricultural extension agents were presented in chapter 2. Only one of these indices captures the dimension of work effort, however. This measure is the average number of farmers that an agent visits on a day devoted wholly to that activity—what we have labeled Visit Effort.

The public show of subservience maintained by the junior staff is such that originally we did not expect to find any work-group restriction of output. We first stumbled on it in examining the various correlates of Visit Effort. Several of the variables related to Visit Effort are ones that would be likely to be associated with it only if work-group restriction were taking place. We shall return to these various correlates shortly, as they provide important clues to the conditions under which locational groups do undertake output restriction. In themselves, however, these related variables do not prove the existence of group control over production, for their evidence is only indirect. It was precisely to establish more conclusively that group restriction exists that we conducted the second set of interviews that we have been discussing in this chapter.

In these follow-up interviews, we began by asking, "When an AA or JAA is visiting farmers, how many should he see in one day in order to be doing his work fairly?" All the men interviewed answered with definite numbers, which range from 1 to 10 and average 4.4. There also are statistically significant differences between the five locations sampled in the numbers given, suggesting group standards on this issue $[F(4/47) = 3.29, p < .05]$.[3] The next question was, "Do you and your fellow workers ever talk about what a fair day's work is?" Only 18 percent (9/49) said No. We followed by asking, first, whether any group pressure is ever put on those who do less than the number of visits they had indicated. Then came, "Do you and your fellows ever criticize a man who does a lot more than this fair amount? Do you try to encourage him not to make the rest of you look bad?" As can be seen in table 13, a majority were willing to admit that they criticize both those fellow workers who do too much and those who do too little work and encourage them to keep to their group's definition of a fair day's work. Given the reluctance of the agents to acknowledge publicly that they ever conflict with those in authority, these admissions of collective control of work output are strong evidence of its existence.

Of course some agents felt they needed to explain why they should discourage "too much" work. They argued that too many farm visits

TABLE 13
Junior Staff Reports of Group Output Control

Number	Percentage	Responses to Questions about Whether the Location Work Group Establishes and Enforces a "Fair Day's Work" Standard
9	18	Never discuss
5	10	Discuss but never criticize
10	20	Criticize too little but not too much work
25	51	Criticize both too little and too much work
49	100[a]	

[a]The apparent total of 99% derives from rounding errors.

mean that the farmer is not being properly taught. At a certain point this is obviously true. One member of senior staff (an AAO) who had done farm visits on foot himself suggested to us that five farmers is the number at which the agent is both keeping busy and doing a reasonable job. Our research, remember, reveals that the actual number of farmers seen on a visiting day averages 1.75, less than half the average "fair day's work" proposed to us by the junior staff. It therefore seems unlikely that much of the control exercised over the "excessive" work of colleagues arises from a desire to improve the overall effectiveness of their extension activities. Most of the output restriction is almost certainly directed at lessening the work load and protecting one another. Because the actual amount of work done is so low relative to the junior staff's own definition of a fair amount, it is doubtful that the work group is very serious about criticism of underproducers.

The majority who admit that the work group is engaged in output restriction were evenly spread throughout the five locations in our subsample. Since this majority was reporting the behavior of their fellow workers and not just of themselves, it is most likely that those who, in varying degrees, did not admit the existence of restriction were actually only unwilling to reveal the group's secrets to an outsider.[4] We believe that the urge for group restriction of output is general. Nonetheless, this basic propensity is not always accomplished very well. Let us now return to our original data on actual Visit Effort and examine the conditions under which the location work group is more or less effective in gaining control over the output of its members.

A check by analysis of variance of whether the work groups are determinants of the Visit Effort of their members shows statistically significant differences between the fourteen locational groups of agricultural agents [$F(13/130) = 2.54$, $p < .01$]. The group is influencing the behavior of its members and thus the characteristics of the group and not of the individual agents need to be studied. First, groups will not be equally successful in controlling the actions of their members. More cohesive groups should have more collective influence. Also, this internal control could be used either to increase or decrease the level of productivity in the group. Thus the research of Stanley Seashore indicates a good deal of variation in output level between high-cohesion groups (because some raise the productivity of their members while others depress it) while low-cohesion groups show relatively little variation between one another (because the individuals in them are not controlled and their differences tend to average out).[5] When tested, our data reveal the same pattern. We separated out the six most cohesive work groups, using as our indicator the average proportion of the group that a member names as his friends (described in chapter 3). The standard deviation in the Visit Effort averages of these six is 0.738, opposed to 0.235 for the low-cohesion ones. Such a difference would occur by chance alone in less than 1 out of 100 samples of the same size $[F(6/8) = 12.7, p < .01]$. When one does not know whether group output restriction is occurring in an organization, Seashore's indirect indicator is very useful. Given some information on the job performance of individuals and some idea of the social cohesion of the work groups, Seashore's test can give an idea of conflict-group behavior.

Another determinant of the work group's ability to control its members' productivity is the complexity of doing so. Under some conditions it is simply too difficult for group members to keep track of the work of one another and to actively encourage keeping a "fair" standard. Agricultural extension activities obviously present major problems of this sort, for the individual agent makes his farm visits completely out of sight of his colleagues and he meets with them only irregularly. Given the basic propensity of junior staff to engage in output restriction, any increases in their possibilities for interaction would lead to improvements in their collective control mechanisms. Thus we find that the Visit Effort of individual agents decreases with the number of locational staff meetings they attend each month ($r = -.20$, $N = 169$, $p < .01$).[6] This correlation is relatively weak and one certainly would not cancel staff meetings because of it. Its importance lies in what it tells us about the general social processes at work. Apparently meetings provide opportunities for extension workers to talk and agree upon restrictions in the amount of work they do. Paradoxically, the very staff meetings called by the supervisors in order to improve their administrative control actually have a weak tendency to achieve the opposite by giving the subordinates a chance to organize. Nonetheless, numbers of group gatherings are only one relatively minor determinant of the interaction process. The sheer complexity of the interpersonal relationships necessary for collective control is even more important.

Over thirty-five years ago Graicunas called attention to the implications of the number of people in a work group for the complexity of the interaction process. Between two people, there is only one social relationship; in a group of six there are fifteen relationships; and a group of twelve has sixty-six potential relationships. The number of possible interpersonal relationships increases much more rapidly than does the number of people involved.[7] Graicunas was concerned about the impossibility of a supervisor's mastering the complexity of the interpersonal relationships in a big group. If large numbers present management problems for a formally sanctioned authority figure, one can well imagine the difficulties they place in the way of a group of subordinates trying to develop an informal organization. As the research of Peter Blau has since proved, size is a basic structural feature with a wide range of implications for an organization's social dynamics.[8] Returning to the Ministry of Agriculture in Western Province (where the work groups range in size from five to eighteen members), we would expect that the smaller the group, the greater its potential effectiveness in output control. If, as we argued earlier, almost all the groups are interested in restricting and decreasing productivity, the smaller groups should have the lower levels of productivity. In fact this is the case; at the group level of analysis, there is a strong positive correlation between Visit Effort and the number of individuals in the location group ($r = .67$, $N = 14$, $p < .005$).[9] Also indicating the impact of size on the effectiveness of group control, there is a tendency for the amount of variance in the Visit Effort of members to be less in the smaller groups (although, perhaps because of the few groups being analyzed and the imperfections in our data, the relationship is not statistically significant: $r = .38$, $N = 14$, $p > .05$).

We have established the existence of two strong influences on the process of group output control. One, group size, is an objective, structural feature, independent of the feelings of its members. The other, group friendship cohesion (the average proportion of other group members named as friends), measures the subjective affections that bind the group members to one another. How do the two variables interact with one another in their influence on output? First, we would expect that the larger groups would have difficulty in achieving friendship cohesion. Our data do show a weak tendency for small groups to be more cohesive, but the relationship is far from being statistically significant ($r = -.22$, N = 14, $p > .05$). Although it may be more difficult, it is possible for the members of large groups to be friends with a fairly large proportion of their fellows.

The most interesting fact about friendship cohesion, however, is that it has quite different effects on output control in large groups than in small ones. We have already mentioned that the small work groups have lower averages on Visit Effort than do the larger ones (1.40 vs. 2.01; see table 14). Increases in friendship cohesion in the small groups (5–9 members) lead to still more restriction of output: the small groups with moderate cohesion have a Visit Effort average of 1.30 opposed to an

TABLE 14
The Size-Cohesion Interaction Effect on the Average Visit Effort of the Location Work Group: The Mean of Group Averages

| Group Size | Group Cohesion | | Mean |
	Low	Moderate	
5–9	1.51	1.30	1.40
10–18	1.88	2.22	2.01
Mean	1.74	1.75	

average of 1.51 for the low cohesion small groups (see table 14). Yet increases in friendship cohesion in the larger groups have exactly the opposite effect. The higher the friendship cohesion in the medium and large groups (10–18 members), the less they restrict output; the larger groups with moderate cohesion register an average Visit Effort of 2.22 against a 1.88 average for the larger low-cohesion groups (see table 14).

The same set of facts is illustrated somewhat differently in table 15. There it can be seen that the six least productive groups have five to nine members and that the eight most productive ones have ten to eighteen in them. The six groups with moderate friendship cohesion include both the three least productive and the two most productive ones. The striking point is that these three lowest output groups are all small and that the two highest output groups are both large. Obviously group cohesion has different implications for output depending on whether the group is small or large. We will call this phenomenon the Size-Cohesion Interaction Effect. Alone it explains 56 percent of the variation in Visit Effort ($r = .75$, N = 14, $p < .01$; with size controlled $p < .05$).[10] How can its effect be explained?

TABLE 15

The Size-Cohesion Interaction Effect on the Average Visit Effort
of the Location Work Group: The Rank Order of Group Averages[a]

| | Group Cohesion | | |
Group Size	Low	Moderate	Totals
5–9	9,10,11	12,13,14	9–14
10–18	3,4,6,7,8	1,2,5	1–8
Totals	3,4,6–11	1,2,5,12–14	

[a] The figures reported in the table are the rank orders of Visit Effort performance of
each of the 14 location work groups. Number 1 represents the most productive work
group, which had an average Visit Effort of 2.49. Number 14 is the group showing the
least Visit Effort, 1.18.

Blau and Scott point out that group cohesion generally raises worker
satisfaction and lowers turnover and absenteeism.[11] The positive
feelings generated by having many friends in one's work group spill over
into positive sentiments toward the place of work. The same effect is
evident for our agricultural work groups. We measured an agent's
morale by combining his answers to three questions, which are highly
intercorrelated: (1) his intention to make the ministry his career, (2) his
willingness to take another, higher paying job if it were offered, and (3)
the intensity of his complaints about the ministry ($r\,1.2 = .48$; $r\,1.3 =
.45$; $r\,2.3 = .43$; $N = 169$, $p < .001$). The group averages of this com-
posite indicator of morale are positively correlated with friendship cohesion ($r = .60$, $N = 14$, $p < .025$). We also would expect men who are
more satisfied to put more effort into their work. A weak relationship
between morale and Visit Effort does exist for individual extension
agents ($r = .18$, $N = 169$, $p < .01$). The argument that friendship
cohesion raises morale, which in turn raises output thus seems a
reasonable one. But it is not universally valid and its applicability in a
given situation depends on other factors.[12]

An alternate and equally reasonable hypothesis would be that
friendship cohesion improves the capacity of the group to exercise
effective control over its members and thus may lead to a lowering of
output.[13] Such an argument would still be able to account for the
better morale of the higher cohesion groups. It is known that workers
take satisfaction in being able to control their work process, and this
would be true even if the control is exercised against management
objectives.[14] For example, active union members in the United States
are likely to be higher in job satisfaction than are their inactive fellow
workers.[15] Thus, one line of reasoning has friendship cohesion raising

group productivity; the other, lowering it. In this particular case, we believe that both arguments are correct and that the one that is in force depends on the size of the work group. We have seen that junior staff have a basic interest in output restriction. The smaller the location group, the more easily an informal organization to counter Ministry goals can be created. When the numbers involved are fewer than ten, the organizational task is simple enough to be commonly attempted, and the more cohesive the group, the more effective it is in controlling its members, thus lowering Visit Effort. With the larger groups (ten to eighteen), informal arrangements are increasingly complex and determined effort is needed to attempt them, whatever their level of effectiveness. In these circumstances a group that is high in friendship cohesion provides a socially satisfying work environment for the extension agents, increases their morale, and saps their resolve to organize for output control, giving it relatively high levels of Visit Effort.

Anonymous Leadership

There are occasions on which the junior staff informally organize for conflict with their superiors, and many location work groups are engaged in collective restriction of farmer visit output. These counter-organizational activities are well hidden by junior staff and are difficult to uncover, but they exist nonetheless. Who are the leaders of these collective actions? We have been unable to find them through statistical analysis. This suggests that the groups probably do not have a stable leadership for conflict situations and that the leaders who emerge from time to time do not have a uniform set of characteristics. Before we discuss the implications of this statement, it would be well to describe the search process we went through in our attempt to locate the leaders of resistance.

Our assumption is that group leaders will be chosen as friends disproportionately often and that if we examine the characteristics of those most often named as friends, we will be able to identify the junior staff's criteria for leaders.[16] Among agricultural extension staff, the strongest correlates of popularity are variables related to formal status. LAAs (the men in charge of the location teams) are most likely to be named as friends ($r = .31$, $N = 169$, $p < .001$); JAAs (those with the lowest status) are least likely to be chosen ($r = -.26$, $N = 169$, $p < .001$). (See also table 9.) (These and other correlations used in this section are not high enough to have much predictive significance, much less to be used directly as a basis for policy decisions. They do provide important clues to the social structure of the work groups, however, and thus are indirectly useful for policy analysis.) The next strongest correlate of popularity is an agent's overall Explanatory Ability. The better able he is to explain why his various technical recommendations are good ones, the more likely he is to be named a friend by his colleagues ($r = .19$, N = 169, $p < 01$).[17] This apparent support for the ministry's goals and hierarchy may be deceptive, however, since there is a general social tendency for people to direct their friendship choices toward those with higher formal status.[18] The LAAs may be looked up to because of the qualifications and authority they hold while other men of lower status actually lead the group's informal activities.

We are looking for informal leaders who can unite the group against the ministry hierarchy. In general, if one is to find such counterleaders, he should exclude from his analysis of friendships all those who hold formal authority, in our case the supervisory AAs.[19] When we do so, we none-theless still find that being named often as a friend is correlated with the degree to which one fulfills the ministry's goals. Junior staff are more likely to name a man as their friend if his Visit Effort is high ($r = .25$, N = 107, $p < .01$), if he is well informed about east coast fever in cattle ($r = .24$, N = 107, $p < .01$), and if he is good at explaining the virtues of the cotton or coffee husbandry practices he recommends ($r = .21$, N = 107, $p < .025$).[20]

In a summary of the American literature, Peter Blau and Richard Scott write,

> It appears that the relationship between informal status and perfor-mance is contingent on work group norms: only if the expert exercise of skills is a dominant value in the group does high status tend to be associated with superior performance and to serve as an incentive promoting it; if the dominant norm standardizes productivity, high status is associated with modal performance.[21]

On the basis of other evidence, we have concluded that the ministry's location work groups believe in the standardization of output, so we would expect the most popular men to be modal producers. When, however, we examine the Visit Effort of the fourteen agents who are the most popular subordinate in each group (that is, not the LAA), we find that they conform to their groups' norms neither more nor less than do the other agents. In fact, the same is true of the LAAs: their Visit Effort scores vary around their respective group means exactly as do those of the other agents.[22] Thus the most popular members of junior staff are not modal producers. Instead popularity tends to go to those who are superior in performance.

Not only have we failed to uncover the leaders of the interest group efforts to standardize production, but we find the junior staff giving social status to those who appear most to defy the group norms on production. The latter phenomenon seems paradoxical, but similar effects have been observed in some United States studies. These investigations have shown that under certain conditions, high informal status can promote resistance to group pressure for conformity. When an individual is well accepted by his group, the group will tolerate his deviation from minor group values; only those who are insecure will feel the need to conform to even the slightest group norms. Still, strong pressure will be put on even the most popular to make their behavior consistent with the group's *central* values, and those who deviate in these areas usually are or will become social outcasts.[23] This analysis suggests that although work group norms in the Ministry of Agriculture promote output restriction, this is not a central value to the junior staff.

There are several reasons that output control is not a central norm for the extension agents. First, not only are they paraprofessional teachers of agriculture, they are almost all farmers themselves. Both roles lead them to admire someone who understands the technicalities of farming well. Second, the junior staff are overwhelmingly local men, and extension work benefits their communities. They find it difficult to attack and ostracize a man who is diligent in helping their neighbors. Third, as extension agents do their work on remote farms, it is very difficult for senior staff to know how much work is being done. Junior staff therefore probably do not feel so threatened by the overproducers among them. Fourth, even the junior staff indirectly acknowledge that they are doing too little work. We noted earlier that junior staff definitions of a fair day's work average out to seeing 4.4 farmers per day, against their actual average of 1.75. (Even if we take junior staff reports of their work at face value and make no discounts for deception, the average is 2.48.) Only a handful of men in the entire province regularly exceed their groups' definitions of a fair day's work. In these circumstances, the cohesive work group can reluctantly admire the hard worker for his energy and skill without feeling threatened by him. Because there is little hierarchical pressure and the actual amount of work done is small, work restriction does not have to be a central value for these groups. The strong work group probably operates to lower production, not so much by attacking the overproducer as by assuring the underproducer that he need not have a guilty conscience and that no one is going to squeal on his laziness. These particular work groups offer their members social protection, guaranteeing collective autonomy rather than imposing much collegial control. Nonetheless, were the level of work and the amount of effective hierarchical pressure to increase dramatically, it would be fairly easy for these groups to strengthen their resistance by engaging in vigorous collegial control. A similar pattern of cohesion and resistance was found in a Nairobi telephone exchange work group by J. Okigo and G. A. Owuor. In their study, not only did the socially integrated group members have a lower level of output than the social outcasts, they also revealed *more* variance in their production levels. The authors conclude that the group was not engaged in social control but was engaged in providing protection against the hierarchy, which otherwise would have put pressure on the low producers.[24]

In all of the social patterns discussed above, the extension agents are exhibiting behavior very similar to that of other highly skilled workers, who likewise are torn between their desire to have collective autonomy and their admiration for the skills of their profession. Medical doctors long protected the inferior practitioners in their group by refusing to testify in malpractice suits and also greatly honored their more out-standing colleagues. We gain in self-esteem by being members of a group both with exceptional skills and with freedom from outside control. The dilemma is created by the frequent tendency of collective autonomy to decrease the pressure felt by group members to maintain the level of their work.

At present informal group efforts at work restriction in the Ministry of Agriculture probably do not have stable leadership. Those agents admired by the junior staff for their training and knowledge often have been co-opted into the authority hierarchy and therefore are reluctant to captain efforts of collective restriction. The other agents lack the ready-made popularity necessary to natural leadership and would have to earn status through outstanding efforts on behalf of the group. Outspoken advocacy of the interests of subordinates is too dangerous to be often attempted, however, for all the junior staff are vulnerable to sanctions from their superiors. Instead, output reduction represents a general mood in the groups which is generated and sustained in a genuinely collegial manner and has only transient leaders of no uniform type. This is why the productivity of the groups is so sensitive to their size and number of friendships. Group action growing out of collegial roots requires a much more tightly knit social organization than do actions led consistently by the same one or two respected individuals. Not only is collective resistance to the ministry's goals carefully hidden; it also appears to have unstable leadership. Hence, we have labeled it anonymous group conflict.

Conclusions

To summarize, not only do socially coherent and self-conscious interest groups exist among the junior staff, they are also engaged in dichotomous conflict with the senior staff. Collective action by subordinates is so carefully hidden, however, that its existence is largely unknown and junior staff are believed to be quite subservient. The covert nature of dichotomous conflict in the ministry is a consequence of the powerful sanctions sometimes used against those who offer open resistance. Interest-group conflict therefore develops in a special pattern, one in which senior staff are left to discover the grievances of their subordinates on their own and in which permanent junior staff interest-group leaders probably do not exist, because of their vulnerability. For the goals of the ministry, the most significant manifestation of this anonymous group conflict is informal restrictions on the amount of work junior staff do. All the subordinate work groups seem to believe in limiting the amount of effort they put into their work, but some are more effective than others in encouraging these standards. Because the junior staff apparently do not have stable informal counterleaders, collective action must usually originate on a peer basis. To be effective, such collegial organization requires a tightly knit group. Consequently, output restriction is greatest in the work groups that are both small and socially cohesive, with size emerging as the more basic variable. The existence of junior staff interest groups that do engage in dichotomous conflict with their superiors has important consequences for the ministry's authority and communications systems. Before we pass on to these topics, we should note that our discoveries give support to the theories of Ralf Dahrendorf: authority is a basic variable in the theory of social conflict, tending to create interest groups and dichotomous patterns of conflict. If interest-group conflicts will form around the hierarchy in the Ministry of Agriculture, where subordinates are widely dispersed, there must be a strong propensity for them to develop in all formal organizations.

What is to be done by extension managers about the informal

organizations of junior staff and collective restrictions on output? The standard responses of Kenyan senior staff has been similar to turn-of-the-century capitalist responses to trade unionism—to try to crush collective organizations by punishing those who show open opposition and by rewarding those subordinate leaders who seem pliable. This response is ineffective. It only drives resistance underground and heightens the alienation of subordinates from their work. Nothing is gained by trying to promote weak interest groups of subordinates. In Western Province even the large location work groups, with poor informal organization, have unacceptably low levels of Visit Effort. Real improvements cannot be achieved by altering the balance of power within the existing system of conflict but only by moving into a new type of relationship.

The early scientific management school attempted to break the hold of the work group by creating incentives for individual productivity among the workers. Frederick Taylor achieved some notable successes with this method, but the human relations school of research ultimately demonstrated that it is based on mistaken premises. Strong work groups make a job more enjoyable for employees involved in them and informal organization can press subordinate staff to raise rather than lower their levels of production. These opportunities to make subordinate interest groups supportive of organizational goals can and should be used.

Our analysis of friendship patterns in the location work groups indicates that junior staff do believe in the worth of extension work and admire those who do it well. These values are probably common to almost all groups of professionals and paraprofessionals. If properly involved in the decision-making processes governing their work, agents will use their informal ties to encourage one another to higher levels of productivity. When subordinate interest groups are encouraged not only at the trade union level but also for the work group, it is possible for grievances to be properly aired, problems negotiated, opinions sought on extension programs, and consent gained for work targets. A full discussion of these issues requires that we also understand other aspects of the organizational system—its formal supervision, incentive, and communications structures. These are the themes of the following chapters. We will return to the role of subordinate interest groups in chapter 10 for a more comprehensive statement. The work group can be co-opted only at the cost of placing some limits on supervisory powers, however. The supportive involvement of work groups would be impossible if we believed that an organization's ability to achieve its goals is determined directly by the amount of power exercised by its authority figures. The invalidity of such a model of authority is the subject of the next chapter.

5 Leadership and .Authority

The Inadequacy of Formal Sanctions

When faced with serious laxness among his subordinates, the natural response of a manager is to overflow with indignation and to threaten punitive action ("If you don't shape up, I'll sack the lot of you"). When denied the free use of his sanctioning power, the manager is likely to feel that he has been deprived of his ability to maintain satisfactory levels of performance among his staff. These responses are grounded in a view of organizational relationships as a zero-sum power contest between superiors and their juniors. ("What I win you lose.") Such a model of organizational behavior may be human but it is so over simple as to be false. Nonetheless, because it is based on supposed common sense, this model is the basis of most practioners' thinking about supervisory relationships.

The power model of authority is much in evidence in Kenya. Most managers have the kind of control over their employees' lives that is thought necessary to make the model work. The formal powers held by senior officers over their extension agents in the Ministry of Agriculture are impressive indeed. Their sanctions are extensive: promotions and increments can be blocked and unpleasant or frequent transfers made almost at will. Dismissals and good promotions are more difficult to obtain for one's subordinates—they take time and determination—but here too, in practice, there is virtually no appeal against supervisory arbitrariness. Similarly, there are few effective limits on what an officer can command a member of junior staff to do. Perhaps there are legal limits on the obligations of government employees, but only the already powerful would dare to seek refuge in the law.[1] We once attended a divisional staff meeting to which the Provincial Director of Agriculture (PDA) had come to promote tea. He announced that all those junior staff who lived in the tea zone and who were not already growing tea would be required to plant it that season. The purchase price of the tea plants would be automatically deducted in advance from their salaries. A brave JAA rose to point out to the PDA that he had a grade milk cow and that his farm was too small to support both tea and a cow. Although it was much more profitable to produce milk than to raise tea and notwithstanding the ministry's encouragement of grade cattle, the PDA directed the man to sell his cow and prepare his land for the tea. The JAA sat down and no further objections were raised. When we talked informally to individual agents afterwards, they expressed no surprise or anger at what had seemed to us an unreasonable and extralegal set of orders. The point of this illustration is not that senior staff make irrational demands of their subordinates—for such occasions actually are rare—but that there are few effective limits on what an officer can command them to do.

In Kenya, belief in the efficacy of the formal sanctions aspect of authority is widespread and deep-seated. Employees are thought to respond only to punishments and wage incentives. The elite conception of the good supervisor is a remote, demanding, and authoritarian figure.[2] This set of attitudes is the more striking because it corresponds closely to that prevalent in the United States before the First World War.[3] Are current Kenyan managers perhaps confronting a type of worker very similar to that which his nineteenth-century American counterpart supervised, a worker with whom authoritarianism is the most effective strategy? Or are both Kenyan and earlier American managers to be judged mistaken?

Equally striking is the acceptance of a similar view of effective supervision among managers in socialist Tanzania. Since 1971 vigorous steps have been taken to stamp out supervisory authoritarianism in Tanzania, but managers have not responded by developing new, socialist styles of supervision. Instead they have retreated to their offices and exercised little leadership at all. The result has been a drastic drop in worker productivity, precisely as the managers had expected. But these expectations were fulfilled because they believed that the only alternative to authoritarianism was withdrawal from supervision altogether. Thus both the practice of authoritarianism in Kenya and its absence in Tanzania beg the same questions: is authoritarian supervision actually productive and are there effective alternatives?

If the formal sanctions conception of effective authority were correct, we should find two important relationships between supervision and work. First, those hierarchical levels that have more control over formal sanctions would have a greater influence on the behavior of subordinates. Second, those characteristics of supervisors that make them more remote and more punitive should produce increased work from their juniors. In fact, neither hypothesized relationship holds in the Kenyan Ministry of Agriculture. Let us examine them in turn.

In chapter 3, we noted that the first-line supervisors—the Location Agricultural Assistants or LAAs—hold little formal authority relative to their senior staff superiors. The only sanctions at the disposal of the LAA are those that he can influence his AAO, the second-line supervisor, to use. The annual staff evaluations on which promotions are based are written by the AAO; only a member of senior staff can place a letter of warning in an employee's file—an action that might lead to his eventual dismissal; only senior officers can command an agent's transfer to another area farther from or nearer to his own farm and home. An LAA's use of these sanctions to support his own authority depends wholly on his influence with his AAO. The formal authority of the LAA is much less than that of the AAO, and on this basis the AAO might be expected to have the more important influence on the work of the junior staff. The opposite is actually true. We took a series of characteristics of the work groups and of their supervisors—the LAAs and AAOs—and examined them through multiple regression analysis in order to discover their relative impact on subordinate staff work behavior. Table 16 presents the resulting regression equation and shows the four variables with the strongest influence on the average Visit Effort of location work groups. The most important determinant of group Visit Effort is the Size-Cohesion Interaction Effect, described in chapter 4. The two factors next in significance are both attributes of the LAA (numbers 2 and 3). The LAA and AAO sets of characteristics that we analyzed were directly comparable and included items such as each superordinate's frequency of supervision, the level of his technical training, and the length of time he had been in charge of this particular work group. The fact that characteristics of the LAA entered the regression equation before the same or any other characteristics of the AAO indicates that the LAA has a more important effect on the Visit Effort of junior staff than does the AAO.[4]

The second hypothesis derived from popular wisdom is that those characteristics of supervisors that make them more remote and more punitive will produce increased work from their juniors. Once more the independent variables in the multiple regression equation predicting Visit Effort lead to a rejection of the hypothesis. The propensity of the LAA to be punitive (LAA Supervisory Effort)[5] leads to a clear reduction, not increase, in Visit Effort. The greater the number of friendships an LAA has with his subordinates (LAA Popularity)—the opposite of

TABLE 16

Multiple Regression Analysis[a] of the Four Variables with the Strongest Effect on the Average Visit Effort of the Location Work Group

Variable Name	Standardized Regression Coefficient[b]	Regression Coefficient	Partial Correlation Coefficient	Variable Mean	Variable Standard Deviation	Variable Minimum	Variable Maximum	T Statistic	Probability of Chance Occurrence[c]
1. Size-cohesion interaction effect[d]	.87	0.46	0.92	-0.05	0.71	-1.47	1.13	7.27	<.001
2. LAA supervisory effort	-.39	-0.22	-0.73	3.21	0.70	2.00	4.00	3.18	<.02
3. LAA popularity	.34	0.61	0.67	0.34	0.21	0.11	0.75	2.72	<.05
4. Luhya AAO	.29	0.23	0.62	0.64	0.50	0.00	1.00	2.40	<.05
(Constant) (Average Visit Effort —the dependent variable)	...	2.35	...	1.75	0.38	1.18	2.49	...	...

Degrees of Freedom = 9

Multiple Correlation Coefficient (R) = 0.94

[a]Multiple regression analysis measures the simultaneous effect of several independent variables upon a single dependent variable. Multiple and partial correlation coefficients take values of 0.00 to 1.00 exactly as does the simple correlation coefficient (see note to table 3). The partial correlation coefficient operates by first removing (controlling) the effect on the dependent variable of all other independent variables in the analysis and then measuring the amount of the remaining variation in the dependent variable that is explained or accounted for by the particular independent variable being examined. The square of the coefficient is the amount of this variation that is explained. The multiple correlation coefficient is the square root of the total amount of dependent variable variation that is explained by all the independent variables

together. In addition, regression analysis provides a formula for predicting the value of the dependent variables from the independent variables. This formula takes the form of $Y = a + \beta_1 x_1 + \beta_2 x_2 \dots \beta_i x_i$, where Y is the dependent variable, the xs are independent variables, betas are the regression coefficients for each x, and alpha is the constant.

[b]The standardized regression coefficient is the regression coefficient (β or beta) that would occur if all independent and dependent variables' means were set at zero and their standard deviations made equal to one. By standardizing in this way, it is possible to compare the relative influence of different variables in an equation—the larger its standardized regression coefficient, the greater its influence on the dependent variable.

[c]All the probabilities of error here are for a two-tailed test of significance. The probabilities would be half for a one-tailed test.

[d]This variable is a product term, i.e., it is derived by multiplying size times friendship cohesion. (The exact way this is done is described in Chap. 4, n. 10.) The question may arise whether the two individual variables—size and cohesion—should also be entered into the equation in addition to the product term. There is no general answer to this question, the appropriate strategy being determined by the explanatory model brought to the analysis. In our case we have no reason to expect either of the two variables to have any influence on Visit Effort other than that which would derive from its interaction with the other. The correctness of this expectation is sustained because neither size nor cohesion can enter the regression equation with statistical significance once the Size-Cohesion Interaction Effect has been included.

remoteness—the better his staff's Visit Effort. In chapter 3, we noted that the Luhya extension agents feel much closer to senior staff of their own tribe than to other officers. Like the LAAs, these less remote Luhya AAOs get more Visit Effort from their subordinates than do the outsider AAOs.

The Kenyan elite are mistaken in their understanding of the dynamics of their supervisory relationships with their staff. Because they have been deceived by the apparent subservience of their subordinates, they have not considered the possibility of collective worker resistance to their authority. The reality described by the regression analysis is consistent with organization theory as it is currently understood in the West.

Authoritarianism vs. Leadership

In *Exchange and Power in Social Life*, Peter Blau points out that

> the distinctive feature of authority is that social norms accepted and enforced by the collectivity of subordinates constrain its individual members to comply with directives of a superior. Compliance is voluntary for the collectivity, but social constraints make it compelling for the individual. In contrast to other forms of influence and power, the pressure to follow suggestions and orders does not come from the superior who gives them but from the collectivity of subordinates. These normative constraints may be institutionalized and pervade the entire society, or they may emerge in a group in social interaction. The latter emergent norms define leadership, which, therefore, is considered a type of authority. The authority in formal organizations entails a combination of institutionalized and leadership elements.[6]

Societal values provide that when a man opts to work for someone, he owes his employer a certain minimum of compliance with his orders. The social pressure which the employee feels to obey these minimum commands provides his superior with formal, institutionalized authority. Societies may define these basic obligations differently, and when peoples from dissimilar cultures first begin to interact, there may be major misunderstandings over the nature of the "employment contract." Early Kenyan white settlers and their African employees frequently found each other irresponsible, in good part because of different cultural perceptions of the meaning of the employment relationship. In most situations, however, and definitely in Kenya today, there is a broad and fairly stable understanding of the minimal obligations entailed by employment.

Blau goes on to argue that

A manager in an organization has some formal authority over subordinates, since they have accepted the contractual obligation to perform the tasks he assigns them, and he has also considerable power over them, since he has official sanctions at his disposal through which he can affect their career chances. The managerial authority that is rooted in the employment contract itself is very limited in scope. It only obligates employees to perform duties assigned to them in accordance with minimum standards. This formal authority does not require them to devote much effort to their work, exercise initiative in carrying it out, or be guided in their performance by the suggestions of superiors. Effective management is impossible within the confines of formal authority alone. A manager may extend his control over subordinates by resorting to his sanctioning power to impose his will on them, promising rewards for conformity and threatening penalties for disobedience. An alternative strategy a manager can use is to provide services to subordinates that obligate them to comply with his directives. His formal power over subordinates helps to create joint obligations, in part simply by refraining from exercising it. In this case, the manager relinquishes some of his official power in exchange for greater legitimate authority over subordinates.[7]

In order to meet his organization's goals, the supervisor can choose either to use his sanctions fully, gaining compliance through the exercise of power, or to be moderate in employing his sanctions, developing a claim to greater legitimate authority and gaining compliance through leadership. The route most often taken by Kenyan managers is power and not leadership.[8]

In principle, power and leadership might be equally effective means of gaining the compliance of subordinates. In practice, leadership generally produces better results, partly because the power alternative fails to take account of the collective counterpower of the workers. Leadership operates by capturing the loyalties of the subordinates, defusing any intentions they may have to engage in collective output restriction and enlisting their social pressures to encourage employees to work. Managerial power, on the other hand, challenges subordinates to try their hand against it. The power position of the manager against any individual member of his staff is generally overwhelming, but against a united and determined collectivity of staff, the balance of power will almost always be reversed.

Once we recognize the existence of the potential for junior staff work-group organization in the Ministry of Agriculture, the attractiveness of the power strategy for gaining compliance fades. This is borne out when we examine what happens to Visit Effort when an LAA does attempt the full exercise of his supervisory powers. We questioned each Location Agricultural Assistant about his supervisory practices. First, we asked how often he inspects the work of the AAs and JAAs under him to make sure they are doing a competent job. All of the LAAs indicated that inspections are a regular part of their work, and nine of the fourteen (64 percent) said they made them frequently. Second, we inquired, "Do you see yourself as *primarily* responsible for supervising and inspecting the JAAs and AAs under you, or for answering the technical questions of the JAAs and AAs, or for working directly with farmers yourself?" Eight of the LAAs (57 percent) gave supervision as their basic job. These men had opted for a taskmaster type relationship with their subordinates rather than one of being helpful or indifferent. Both dimensions of the taskmaster role are threatening to the lax subordinate. LAAs who inspect frequently increase the danger of the subordinate's being exposed; supervisors oriented toward formal authority are more likely to invoke sanctions when laxity is found. Nonetheless, these two dimensions of the authoritarian role are not related to one another from the viewpoint of the supervisor and are completely uncorrelated ($r = -.04$; N = 14). The type of relationship an LAA chooses to have with his subordinates seems to have no influence on how often he goes into the field to observe their work. Thus, the two variables have an additive effect in the way the LAA confronts his extension agents. The most threatening LAAs would be the five who both define their relationship with their juniors as supervisory and who carry out frequent inspections. The least threatening LAAs would be the two who neither see inspection as their primary task nor visit their staff often. In between would be those who have a taskmaster type relationship but do not make frequent inspections and those who basically try to be helpful and who visit their agents often. This combined index of LAA Supervisory Effort is clearly negatively correlated with Visit Effort. The more threatening the LAA is, the less work his subordinates do.[9] Furthermore, although both the kind and amount of supervision are important, it is the taskmaster type of relationship with subordinates that reduces their Visit Effort more.[10]

The Supervisory Effort of the LAA has a negative effect on his group's Visit Effort because his power position is weak. The extension agents work over a wide geographical area. Without sophisticated management methods, such as those we will discuss in chapter 10, it is impossible for the LAAs to monitor more than a tiny fraction of their subordinates' work. If the junior staff wish to deceive their supervisors and do less work, it is not hard for them to do so. If a cohesive work group finds an LAA unpleasant, it is easy for it to defeat his efforts. The ministry's sanctioning powers are simply ineffective for controlling the day-to-day work of its staff.

The alternative available to the LAA is to capture the allegiance of his work group and to lead it to greater effort. If he can do this, it will be an advantage rather than a threat to him to have a more cohesive group of juniors. When set against its supervisor, the stronger work group can more easily defeat him, but when it is supporting him, it offers him more legitimate authority and greater capacity for leadership.

The acceptance and approval of subordinates is gained by exceeding their expectations for their relationship with a supervisor. Peter Blau observes that

> The satisfactions human beings experience in their social associations depend on the expectations they bring to them as well as on the actual benefits they receive in them. The man who expects much from his associates is more easily disappointed in them than the one who expects little, and the same degree of friendliness might attract the first man to other people and discourage the second from associating with them. These expectations of social rewards, in turn, are based on the past social experience of individuals and on the reference standards they have acquired, partly as the result of the benefits they themselves have obtained in the past and partly as the result of learning what benefits others in comparable situations obtain.[11]

Thus the expectations held by subordinates are relative and not absolute. The supervisor does not have to meet some universal standard of behavior in order to get approval, but only to surpass that of other supervisors in the society and in similar organizational positions. By providing services that other supervisors usually do not, by refraining from the use of sanctions that others generally employ, by not enforcing rules that others customarily insist on, the supervisor builds up a sense of social obligation among his subordinates and wins their acceptance.[12] Thus an LAA exceeds the expectations of his juniors if he is helpful toward them while most Kenyan supervisors play the taskmaster and if he shows faith in their dedication to work while most of his counterparts are checking them whenever possible. By pulling back from the frequent harshness of supervision in Kenya, the LAA gains approval from his subordinates, more legitimate authority, and the ability to lead them in greater effort. As in the United States, the relatively more authoritarian supervisor is generally the less effective.[13] The fragments of evidence available from other Kenyan studies likewise indicate that superiors who are nonauthoritarian leaders are usually more successful than those who practice authoritarian domination or pay no attention at all.[14] Similar findings have been reported for India.[15] Thus we seem to have a proposition in organization theory with clear cross-cultural applicability: supervisors in utilitarian organizations who use relatively nonauthoritarian leadership techniques generally achieve higher productivity from their work groups.

Nonetheless, there is still a cultural context to the supervisory relationship. After studying supervisory styles in a rural Kenyan soap factory, Makhanu and Ole concluded that,

> What the Americans see as authoritarian is not likely to be viewed as such by the [Kenyan] African workers. . . . It is an inescapable fact that the American workers have been reared in democratic norms and that they are, comparatively speaking, highly politicized. It seems to us that it will really require a very bad supervisor, who goes far beyond his formal exercise of authority, for an African worker to refer to him as "authoritarian."[16]

It is true that the Kenyan worker is more subservient than his European or American counterparts, that he will endure supervisory practices that would lead them to quit, and that he has much lower expectations of those who are in authority over him. In a Nairobi study, F. K. Muthaura found no relationship between staff turnover and authoritarian supervision. Jobs were simply too scarce for employees to resign because of unpleasant treatment by their superior.[17] But the willingness of Kenyan employees to accept greater amounts of authoritarianism than Europeans in no way contradicts the proposition that they work better for less authoritarian supervisors. The differences in response of Kenyan and, say, American workers to supervisory styles are illustrated in Figure 2. The heavier a superior's reliance upon his sanctioning power the more authoritarian he is. Point 1 represents the amount of authoritarianism that an American employee expects or accepts as tolerable in a supervisor. Point 2 indicates the expected or acceptable

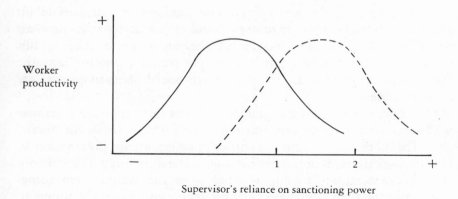

Supervisor's reliance on sanctioning power

Key:
_____ United States
-------- Kenya
1. Expected supervisory style in the United States
2. Expected supervisory style in Kenya

Fig. 2. The relationship between supervisory style and productivity in the United States and Kenya.

level of authoritarianism for Kenyans. The average Kenyan finds acceptable more authoritarianism than does the average American. Nevertheless, relative to their different expectations, both sets of workers respond in the same way. Supervisors who are more authoritarian than their employees think acceptable find productivity dropping off very rapidly. Superiors who rely upon their sanctioning power somewhat less than their juniors expect create social obligations toward them among their subordinates, have leadership, and attain the highest levels of productivity. Workers under those who use sanctions much less than they expect, however, sense that their superiors are indifferent rather then kind, feel no obligation toward them, and produce the least of all. The American literature terms the left-hand third of the United States curve laissez-faire supervision—the least productive; the middle third is labeled helpful or democratic supervision—the most productive; and the right-hand third is called authoritarian—medium in productivity.[18] If we define *authoritarianism* in absolute terms, then somewhat authoritarian supervision in the United States is the most effective superv:sory style for Kenyan conditions. If, however, authoritarianism is defined relative to the culture's own expectations of supervisory behavior, it is always inferior to a helpful style. It is important to keep both of these aspects of the supervisory style–productivity relationship in mind. Nonetheless, if we are to understand the social dynamics of what is happening, it is the behavior of the supervisor relative to his subordinates' expectations that is basic. The superior who is relatively, not absolutely, good is the one who creates social obligations and can exercise leadership.[19]

Of course, the supervisor applies and refrains from applying sanctions within the context of an organization, which greatly limits his discretion. The more extreme forms of authoritarianism are simply impossible in some organizations because there is no hierarchical support for them. In other organizations, the higher echelons are committed to rewarding and punishing their employees, and no matter how much the immediate supervisor may refrain from using his personal sanctioning power, he can never create a sanctionless or laissez-faire atmosphere for his subordinates. This latter situation is the sort faced by the LAA in the Ministry of Agriculture. The senior officers are adamant that their orders be obeyed and committed to punishing those that defy them. Since the work is geographically so dispersed, the officers' judgments of their subordinates are often defective and sometimes overpersonalized.[20] Nonetheless, the atmosphere is already charged with sanctions and there is nothing the LAA can do to make it truly laissez-faire. Thus a nonthreatening LAA seldom runs the risk of lowering productivity by making the total supervisory context too lax. The authoritarian LAA can make the context considerably more threatening and sanction-ridden, however, by inspecting and reporting on his AAs and JAAs. Hence, in the Ministry of Agriculture, it is the nonthreatening LAA who builds up social obligations and gets better Visit Effort from his staff. The threatening LAA is seen as overly authoritarian and brings on a negative group reaction and decline in Visit Effort.

The argument made in this section has taken account of differences in national culture, but it applies to intracultural differences as well. Obviously there are differences in the expectations of workers within, as well as between, cultures. In addition to purely individual sources of variation in expectations, there are those associated with social classes, kinds of work, and types of organizations. Both in the United States and in Kenya, for example, managerial level employees expect less frequent supervision than do those at the very bottom of the authority pyramid.[21] They also expect that the supervision they do receive will be considerably less authoritarian than that which they mete out to their subordinates. Here too the effective supervisor is one who is seen as good relative to these expectations and who creates social obligations toward him among his managerial staff.

Three other Kenyan case studies add weight to our argument and illustrate our analysis further. The first is a comparative analysis of the effects of supervisory style in twenty rural coffee factories by C. M. Ambutu and six other students at the University of Nairobi. The work done in the factories is not complicated and the employees tend to be poorly paid, functionally illiterate, older men. The authoritarian factory managers are able to get their staff to process larger quantities of coffee per man month, generally by forcing them to work overtime without pay. But the result is that the workers become careless in their handling of the crop. The nonauthoritarian managers generate a greater sense of responsibility among their subordinates, get better quality coffee beans as a result, and make at least 15 percent higher net profits for their factories.[22]

R. H. O. Kowitti's study in the Kisumu Cotton Mills demonstrates the differences in effectiveness of laissez-faire and authoritarian supervisors. During the three-month period which he studied, the twenty-five weavers under a laissez-faire jobber had a consistently lower level of production than did the fifty under the two authoritarian jobbers. Kowitti did not find any nonauthoritarian supervisors in the American sense. He notes, however, that his two authoritarian jobbers, though strict, worked closely with their weavers and succeeded in getting them bonuses and promotions. They provided their subordinates with valued services and Kowitti found that this was reciprocated with strong loyalty. Clearly it is the exchange of services rather than their particular form which is most important.[23]

The previous two cases concern relatively unskilled manual laborers. In neither case did the production process allow any meaningful opportunities for involving the workers in regular decision making about their jobs. Such participation did not seem to be high on the list of priorities for these employees and its absence did not bother them. Even the coffee factory managers who were called nonauthoritarian by their subordinates did not consult them when making job-related decisions. The situation is quite different in the third case available to us—an examination of the relationships between headmasters and teachers in eleven East African secondary schools. E. K. Arap-Bii and his six colleagues found consistent evidence that helpful headmasters are more effective than authoritarian ones on nine different performance indicators (though owing to the small sample the results were not statistically significant). More striking, however, was a very strong (and significant) relationship between the frequency with which the headmaster reports consulting his teachers, on one hand, and the quality of their performance and their relationships with him, on the other. Clearly secondary school teachers in Kenya (unlike manual laborers) expect to be involved in making decisions about their work and the issue is significant to them.[24] Employees at this level are implicitly prepared to exchange better performance for their participation. The general point is that expectations are formed not only by the national setting but also by social classes, kinds of work, and types of organization. The supervisor who exceeds his subordinates' expectations—whatever they are—will attract their loyalties and be able to ask for better performance in return.

Leader Effectiveness

The effectiveness of leadership depends on the social support of subordinates. The summary term authoritarian has been used above to describe the supervisory characteristics that often cause subordinates to withdraw this needed support. One attribute of authoritarian supervisors is that they do not consult their workers on matters concerning them, but they have a number of other important characteristics as well. In their study of twenty coffee factories, Ambutu et al. asked the subordinate employees a number of questions about their managers in order to see which characteristics combine to form an authoritarian syndrome. Comparing their findings with those of Blau and Scott in a U.S. social work agency, they report that

> managers that workers consider authoritarian also are said to supervise closely, to be procedure-oriented, to be unfair, and to lose their tempers. In addition to the characteristics suggested by the American research, authoritarian supervisors were seen as lacking in self-confidence and as not being very competent in their work.[25]

This suggests that many supervisors resort to their formal powers because of their insecurity and because of difficulty in commanding the respect of their subordinates. The manager who is tempted to increase his use of sanctions might ask himself if he needs instead to demonstrate his own technical competence.

As a general rule, the social approval of subordinates derives from the supervisor's having personal characteristics valued by the group or from his helping the members of the group to obtain something of value to them. Returning to agricultural extension, one of the variables that raises Visit Effort and that is an indicator of subordinate support is LAA Popularity—the proportion of the work-group members who name the LAA as their friend (see table 16). Naming someone as a friend is a clear statement of social acceptance and an indication of a relationship to the other that involves at least some willingness to comply with his wishes. On average, LAAs are named as friends by a third of their staff, with a range of 11 to 75 percent. What causes junior staff to regard some LAAs as their friends more often than others? In the first place, these particular supervisors are probably friendly themselves—they encourage and enjoy informal personal interaction, which their subordinates value. This characteristic removes some of the formality and threat from the authority relationship and makes it more pleasant. Thus Blau and Scott report from the United States that unfriendliness is usually associated with authoritarian supervision.[26] Presumably most of the popular LAAs are friendly men, and the positive feelings thereby engendered among their subordinates increases their standing as leaders.

If friendliness passes over from a generalized relationship with all one's subordinates to strong specific ties with individual juniors, however, in some circumstances it can damage the supervisor's leadership status. Specific friendships create the possibility of favoritism and unfairness if the supervisor has the ability to dispense rewards. The danger of favoritism seems to be felt with particular acuteness in Kenya. LAAs do not control organizational rewards, so their friendships pose no threat to those who are not close to them. The situation is quite different with the AAO, however, for rewards are part of his powers. The positive feelings he generates by being open to friendships with his subordinates are balanced by fears that these will lead to favoritism. Thus the number of times an AAO is named a friend by his junior staff members is quite unrelated to their Visit Effort. Similarly, in the study of Kenyan coffee factories, the friendliness of the manager was completely uncorrelated with productivity and was not related to the propensity of his workers to call him authoritarian.[27] At least in Kenya, a superior with power has to be careful how he expresses his friendliness in order not to jeopardize his reputation for fairness.

There are other things besides friendliness that make an LAA popular. Needless to say, the longer an LAA has been in charge of a particular group, the more opportunity he will have had to build up his social relationships and the larger the number who will think of him as a friend ($r = .63$, $N = 14$, $p < .01$). Further, there is a tendency for junior staff to prefer the LAAs who are most like themselves and with whom they feel more comfortable. Thus we find a weak negative relationship between LAA Popularity and both how much standard education he has and the amount of formal training in agriculture he has received ($r = -.42$, $r = -.40$, respectively; $N = 14$, $p < .10$).

One of the more interesting effects upon the popularity of the LAA is the character of his own superior, the AAO. Junior staff support for the LAA rises and falls in response to their perceived need for his protection from the AAO. LAA Popularity gains as the frequency of the AAO's staff inspections increases ($r = .46$, $N = 14$, $p < .05$). As more junior staff name the AAO as a friend, there is some tendency for less to claim friendship with the LAA ($r = -.39$, $N = 14$, $p < .10$). When the AAO is new to the area and an unknown threat to all, LAA Popularity has a weak propensity to jump ($r = .35$, $N = 14$, $p > .10$). Finally, when the AAO is a non-Luhya, an outsider, there is a very weak tendency for the LAA to receive more friendship choices ($r = .28$, $N = 14$, $p > .10$). All of these correlations are consistent with the theory of social relationships we have advanced here. Social support is given in an exchange relationship, in return for benefits received or expected. AAs and JAAs give support to a nonauthoritarian LAA in appreciation for the sanctions he does not apply.

In the present context, the junior staff give the social approval of friendship to their LAA when they need his protection from a threatening AAO or when he seems to them to be a better man than his own superior. As Blau and Scott first hypothesized, there is a tendency for subordinate loyalty to be expressed on alternate levels of the organizational hierarchy. When workers give their support to their first-line supervisor, they and he gain a mutual strength that enables them both to be independent of the second-line superior. Conversely, when the workers are disloyal to their first-line supervisor, they make him dependent for support on his own superior and this second-line supervisor tends to be the focus of worker loyalty instead. Thus supervisors find themselves in competition with one another, even though it is often weak and hidden. The supervisor who can gain the support of his subordinates will be able to assert some independence with his own superior. On the other hand, the second-line supervisor who can capture the affections of the workers from the first-line supervisors beneath him assures the dependence of both groups of subordinates and strengthens his own position.[28]

Which of them "wins," of course, affects the goal achievement of the organization. As to Visit Effort in the Ministry of Agriculture, the LAA is the more important figure. The AAO who values achievement in this area would therefore never want to purchase the affection of his agents at the expense of his first-line supervisors. If competition did emerge between them, he would even want his LAAs to win the loyalty of the junior staff away from him. Strong ties are needed between first-line supervisors and their staff, and the second-line supervisor who builds his own popularity by weakening these ties will ultimately damage the organization's ability to perform well.

The Second-Line Supervisor

We need to look further at the authority of the AAO, the second-line supervisor. Our attention has been concentrated on the factors that enhance the leadership or legitimate authority of the LAA, who is the first line of authority. This is because the LAA is the one who has the most variable effect upon junior staff Visit Effort. Within the present framework of organizational relationships in the ministry, the LAA is in the most strategic position for adding small increments to agent effort. Yet, outside the spotlight of our attention, the AAO exerts a powerful influence on the LAA. The importance of the AAO is not so readily discerned in the type of analysis we have been making, because by and large it is a constant. The AAO represents the unvarying, forbidding presence of ministry sanctions. Without them, the whole pattern of organizational relationships would be different.

Some AAOs try harder than others to supervise the efforts of their extension agents. As we shall see in chapter 10, the managerial tools they are using are so inadequate that their attempts are without much effect. We later will suggest ways to improve these tools but the basic variable influence on Visit Effort will continue to derive from the leadership of the LAAs. We have already noted that the AAOs, unlike the LAAs, gain no additional social support and legitimate authority from being friendly with their subordinates. Since they control the power to reward, the fear of favoritism balances out the positive appreciation of accessibility.

Nevertheless, there is one characteristic of AAOs that makes them more approachable without suggesting favoritism—belonging to the same tribe as their subordinates. In Western Province, virtually all the junior staff are Luhyas, generally working in or very near their home areas. Forty-four percent of the senior staff in the province and 36 percent of the AAOs in charge of divisions are also Luhyas. It has been the ministry's practice not to post officers to their home areas. Hence the Luhya AAO in charge of a division belongs to the same tribe as his subordinates but he is not a member of their clan or lineage groups. Given the larger context of ethnic politics in Kenya, such a Luhya AAO would be expected to do more for his staff and the division than would an outsider but he would not be suspected of any "unfair" attachments to any of the subgroups in the area. The local extension agents believe that a local AAO is likely to be pursuing more vigorously the agricultural development of their home area, that he will argue their cases more forcefully on matters of promotion and special training courses, and that they can expect a more sympathetic hearing from him when they approach him with their problems. Because he is not from the immediate area, the local AAO also carries no special dangers of favoritism. In sum, the extension agents find the local or Luhya AAO to be deserving of their support, a man whose leadership it is desirable to follow. The results of their legitimation of his authority are evident in their increased Visit Effort.

Summary

In conclusion, we have accepted Peter Blau's social exchange theory of authority. Those in positions of formal authority who provide their subordinates with more benefits (or fewer deficits) than they expected win social approval and support from them. This support can be transformed into greater legitimate authority (or leadership) and used to pursue the authority's goals. Thus we found that the LAA who is relatively nonauthoritarian and the one who is popular with his subordinates is able to capture the support of his work group and get more Visit Effort from it. Similarly, the AAO who, by virtue of belonging to the same tribe as his subordinates, seems more approachable and supportive to them, gains in legitimate authority, and obtains more Visit Effort from them. Some of the details of these leadership-creating processes are non-Western, such as the expectations of relatively authoritarian supervisory behavior and the attachment to tribe. But the underlying social dynamics by which a supervisor turns formal authority into leadership are the same as those in Western theory and seem to reflect a universal aspect of organizational relationships.

Part 3

Organizational
Structures
and the
Performance
of the
Individual
Agent

6 Motivation and Overeducation

Education and Output

Men respond to their organizational environment both as collectivities and as individuals. Our concern in the previous three chapters has been with the degree to which agricultural extension agents are an interest group, their formation into work groups, and the effect of these collective entities on their organizational behavior. We turn now to the interaction between the organization and the individual worker.

The most persistent theme in commentaries on administration in developing nations is the shortage of skilled manpower. A large proportion of the government personnel in poor countries have lower educational and technical qualifications than their counterparts in the industrialized world and it has been assumed that this deficiency accounts for their performing less well. Very large amounts of money and effort have been invested in general and technical education in the last twenty-five years in order to attack this problem, leading to a very considerable rise in the skill levels of government personnel throughout the developing world. Administrative performance has not improved to anything like the degree expected, however. The rapid escalation of personnel qualifications has in turn created new problems. If the new human skill resources are to be used to full benefit, these problems must be understood and organizational strategies developed to deal with them.

Consistent with the general pattern in the developing world, it is the formal qualifications of potential recruits that concern Kenyan employers most. The Ministry of Agriculture recognizes that qualities such as intelligence, motivation, academic preparation, and farm experience are all important to good extension work. Yet the ease of ascertaining standard educational achievement and formal training in agriculture has tended to force out of consideration those candidates who are less qualified than their competitors in these areas but who may be outstanding by other criteria. It is simply assumed that these formal qualifications represent employee capabilities that are good in themselves and that, at worst, will have no negative effect on the other qualities being sought. Focusing only upon standard education, we find that the predominance of formal criteria of selection and the increasing supply of secondary school leavers in Kenya has led to a clear rise in the educational level of new extension workers. [A school leaver is someone who has stopped (left) school at a particular level of instruction, usually to enter the job market.] Those entering certificate courses in agriculture and becoming AAs will now need a good (secondary) School Certificate, where before Independence a good primary school certificate would have sufficed (that is, seven vs. eleven years of education).[1] Even those hired directly as untrained JAAs seem generally to have a Kenya Junior Secondary Examination pass now. Before Uhuru (independence), most JAAs would have been Standard IV leavers (would have had four as opposed to the present nine years of schooling).

Contrary to the expectations of the ministry's recruiters, the rise in the educational qualifications of agricultural extension agents has had a detrimental effect on their performance. The greatest Visit Effort is made by those who have not attained their Certificate of Primary Education (that is, who have had one to six years of schooling).[2] When we turn to Agricultural Informedness and Explanatory Ability, those with upper primary education (Standard V through Certificate of Primary Education) have a clear tendency to know more and explain better than those with either secondary education or only lower primary schooling.[3] This curvilinear tendency is evident in all technical areas tested and persists when we control for other determinants of extension worker informedness (see tables 21 and 22). The general pattern is well illustrated in table 17. We do find evidence that the more complicated the skill, the higher the educational level at which the best performance is usually achieved. On Visit Effort, peak performance is achieved at about five years of education; on Agricultural Informedness, at six years; on Agricultural Explanatory Ability, between six and seven years; and on Innovativeness, at seven years.[4] Thus education is not irrelevant to extension work; those who are best at the most intellectually demanding skills do have the Certificate of Primary Education. Yet the attainment of a secondary school qualification leads to a clear drop in performance in all the areas examined. In every aspect of extension work, the output of agents with secondary schooling is inferior to that of those who have been only to primary school. These are not isolated findings, for R. K. Harrison has found precisely the same phenomenon among extension agents in Western Nigeria.[5] Something happens in conjunction with higher levels of education that lowers the individual's likelihood of doing good extension work.

TABLE 17
Total Information Scores on Hybrid Corn Fertilizers
According to the Education of Junior Agricultural Extension Staff

Education	Mean Score	Variance	Number of Cases
Standards I–IV (1–4 years)	11.2	17.72	14
Between Standard IV and C.P.E. (5–6.5 years)	14.2	9.64	73
Certificate of Primary Education (C.P.E.; 7 years)	14.2	13.75	46
Kenya Junior Secondary Examination (9 years)	13.1	14.86	16
School Certificate (11 years)	12.8	11.36	19

$F(5/163) = 2.37^a$
$p < .05$

[a] The method of statistical analysis used here is called Analysis of Variance. The technique is designed to assess whether the various differences in dependent variable means for all the categories of the independent variable could have occurred by chance alone. The assessment is made through an analysis of the variances (standard deviations squared) in the dependent variable for the various categories of the independent variable and the population as a whole. An F-statistic is used to judge the probability that the mean differences occurred by chance alone. Once one has been assured that the probability of chance error is sufficiently small ($p < .05$), a table such as this one should be studied by looking at the means associated with each of the categories.

The Kenyan educational system consists of a pyramid of school places narrowing very rapidly after the primary level. Those who gain access to each succeeding stage do so on the basis of highly competitive national examinations which test the recall of set bodies of knowledge. It is very unlikely that agents who have been to secondary school and thus have been quite successful competitors in this system have less capacity to learn and understand new information than do the primary-educated junior staff. It is the individual's motivation to do good agricultural extension work that seems detrimentally affected by overeducation. As Harrison states,

there is no evidence to suggest that unusual abilities deriving from intellectual capacity are required for village level extension work; on the contrary, it would seem that the abilities required are possessed by many people *and are brought out* by the greater motivation which manifests itself in better work behaviour.[6]

Just how then does secondary education lead to a decline in extension agent motivation? The rest of this chapter deals with the answer to this question and with ways to overcome the problem of overeducation. In the process of dealing with these specific motivational problems, we will introduce several general theories of motivation. Our objective is to provide a framework within which to understand motivation in the organizations of developing countries. This will then enable us to move beyond the problem of overeducation itself to the more general issue of incentive structures.

The Motivation to Participate

The processes by which work performance is motivated are as complex as the human personality itself. American students of organizations have advanced several theories of motivation, with no accepted resolution between them.[7] We will therefore approach the problem of overeducation and motivation among extension agents by examining several theories in order to find explanations that fit well with our data. We hope that such an analysis will lead to some advancement of our understanding of the factors motivating lower and middle level Kenyan employees.

James March and Herbert Simon have argued that the issue of employee motivation must be divided into two quite different types of decisions:

> The first is the decision to participate in the organization—or to leave the organization. The second is the decision to produce or to refuse to produce at the rate demanded by the organizational hierarchy. The production decision is substantially different from the participation decision in that it evokes a significantly different set [of considerations]. At least some of the confusion in the literature on morale and satisfaction stems from a failure to distinguish between turnover and productivity.[8]

As Victor Vroom makes clear in his important study, *Work and Motivation*, we have much clearer knowledge of the determinants of turnover than of productivity.[9] Since we understand better the decision to continue to participate in the organization, it may help us to begin with that aspect of motivation. It is true that turnover is not a real problem with junior extension staff. Thus a number of the correlation coefficients on which we report are much lower than they would be in a situation where turnover was significant. Nonetheless an analysis of the agent's decision to participate will improve our general understanding of the determinants of turnover. These are important in the case of senior staff, where attrition is more of a problem. Furthermore, in the process we will be able to review several factors that are also important to the worker's motivation to produce.

March and Simon present a model of the resignation decision that is reasonably straightforward, that brings together much of what is found in the empirical literature, and that fits our data well. Even though certain aspects of the model could in principle be applied to any type of organization, March and Simon clearly have developed it primarily for organizations which basically rely on material incentives. Of course our study is concerned with such a utilitarian organization. The elements of the model particularly relevant to the Ministry of Agriculture are set forth in figure 3 exactly as they are specified by March and Simon. One of the advantages of their book, *Organizations*, is that its hypotheses are sufficiently explicit to be expressed in just such a network of causal relationships.

The model is based on what the authors call the Bernard-Simon theory of organizational equilibrium. This theory is really only a working assumption, true by definition, but it does provide an ordering principle which enables March and Simon to give a logical structure to their analysis. The theory is that, "each participant will continue his participation in an organization only so long as the inducements offered him are as great or greater (measured in terms of *his* values and in terms of the alternatives open to him) than the contributions he is asked to make." [10] All work is assumed to have both benefits and costs to the employee. He is asked to do and put up with many things he dislikes. On the other hand, the job provides benefits and may have aspects which the worker finds satisfying. One can imagine these debits and credits of the job being entered in a mental set of accounts and a personal balance of satisfactions being calculated. The individual will leave his present employment when he is convinced that another alternative is actually open to him that will provide a better balance of all his sources of satisfaction (money, social relationships, fulfillment of moral beliefs, and the like) (4.2 : -4.1). [11]

Whether the employee's satisfaction balance is positive (in favor of continuing) or negative (leading to resignation) depends on what he

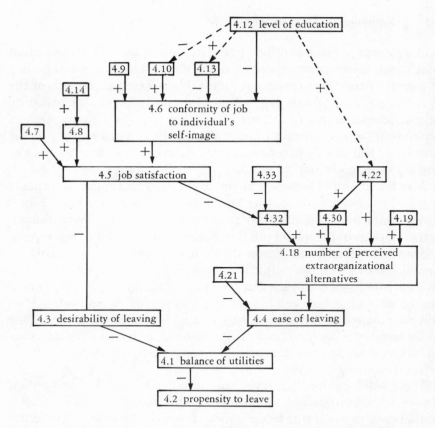

Fig. 3. Portions of the March and Simon model of the job resignation decision

Key:

_____ Causal relationships as hypothesized by March and Simon, showing direction of causation

-------- Additional causal relationships as hypothesized by Leonard

  +   Positive correlation

  –   Negative correlation

4.1 Balance of inducement utilities over contribution utilities

4.2 Propensity of individual participant to leave

4.3 Perceived desirability of leaving the organization

4.4 Perceived ease of movement from the organization

4.5 Satisfaction with the job

4.6 Conformity of the job characteristics to the self-characterization held by the individual

4.7 Predictability of instrumental relationships on the job

4.8 Compatibility of work requirements with the requirements of other roles

4.9 Consistency of supervisory practices with employee independence

4.10 Amount of rewards

4.12 Level of education

4.13 Rate of change of status and/or income in the past (raises and promotions)
4.14 Congruence of work time patterns with those of other roles
4.18 Number of perceived extraorganizational alternatives
4.19 Level of business activity
4.21 Age of participant
4.22 Social status of participant
4.30 Visibility of an individual
4.32 Individual's propensity to search
4.33 Habituation to a particular job or organization
SOURCE: March and Simon, *Organizations*, pp. 93–106 (their numbering).

thinks is (1) the desirability of leaving and (2) the ease of leaving (4.1: -4.3; 4.1: -4.4). Both components are important. The desire to leave the organization does not lead automatically to a resignation. It also must be practically possible to leave. The major determinant of the ease of leaving is whether alternative jobs are available (4.4: +4.18). These points are clearly brought out in our data. We asked each agent, "As you see it at present, do you intend to make working for the Ministry of Agriculture your career?" The answers reveal a very strong expectation of staying with the ministry. On a five-point scale ranging from 0 (for a definite decision to leave) to 4 (for a clear plan to stay indefinitely), the average agent score is 3.5. Some of the agents' commitments to the ministry are based on the correct belief that other good jobs are not available to them. Thus when we asked, "If another somewhat better paying job were offered to you outside the ministry, would you take it?" their commitment score drops from 3.5 to 2.8. An additional factor affecting the employee's perception of the ease of changing to another job is his age (4.4: -4.21). The older he gets, the more interest he has built up in a pension plan and the harder it is for him to learn a new job. Thus we find that the older an extension agent is, the less likely he is to be interested in new employment ($r = -.20$, N = 169, $p < .01$). Even when we have discounted for the difficulty of going elsewhere, however, the extension staff still seem to find the ministry a generally desirable place to work.

The wish to leave a particular employer is basically determined by how dissatisfied the employee is with his job (4.3: -4.5). We asked each agent what complaints he had about his work and then rated his degree of satisfaction on the basis of the number and intensity of his grievances. In general, the junior staff were reasonably satisfied, averaging 2.7 on a scale ranging from 1 (very dissatisfied) to 4 (quite satisfied). There is a clear relationship between an agent's job satisfaction and his commitment to the ministry for a career ($r = .44$, N = 169, $p < .001$). Since job satisfaction is only one of the two determinants of the decision to participate (the other being the ease of leaving), the relationship is a moderate and not a strong one.

What is so satisfying about extension work? First, there is the relative ease with which the agent can do his job and meet his other social and economic responsibilities as well. When the times and places of work are such as to facilitate the worker's meeting his other social role requirements, his satisfaction will be increased (4.5: +4.8: +4.14). The great majority of junior staff are able to live at home and work flexible hours at a relaxed pace. Most middle and low income employees in Kenya have to work in the cities while their families live on the family plot in the rural areas, to be visited only on long holidays. Virtually all wage earners have to manage their farms *in absentia*. The extension agent has the relatively rare opportunity to be a full-time father and farm decision maker while also drawing a salary. Furthermore, the demands of the job are such that he can easily squeeze in extra hours working on his own farm or chatting with friends in the market place. With such an easy congruence between the demands of his several social roles, there is no wonder that agents find extension a satisfying job.

A second part of the desirability of employment with the ministry is that it is secure. The more predictable the vital aspects of a job are, the more satisfying it is likely to be (4.5: +4.7). One of the much remarked advantages of civil service employment is that the worker is certain not to be dismissed as long as he is cautious and performs at a minimal level of competence. Thus a study in Ghana found that fully 74 percent of the civil servants mention security as the reason they like government employment.[12]

The third and, to our thinking, the most important determinant of job satisfaction is the degree to which the characteristics of the work conform to the individual's image of himself in his working role (4.4: +4.6). In the worker's own mind, is the job worthy of him or is it beneath him? This is precisely the point at which we begin to see the effects of overeducation on extension agent satisfaction. March and Simon point out that the larger the rewards in money and status provided by a job, the more worthy and satisfying it appears (4.5: +4.6: +4.10). As it happens, a high salary is not only financially rewarding but is also a symbol of a high social status. Given their educational qualifications, junior staff in the ministry are paid above-average salaries, even for the modern sector of the economy (see table

5). They also belong to the middle elite in the rural areas, holding a status only a bit lower than that of a school teacher, a highly respected position (see table 7). The junior staff are well rewarded, but relatively speaking, they are less well rewarded the higher their educational qualifications. An AA with School Certificate (eleven years) and a Certificate in Agriculture (two years) starts in the ministry at 615 shillings ($88) per month, about 20 percent above what he would be offered in the private sector. Yet a JAA with nothing but his Certificate of Primary Education (seven years) begins at 333 shillings ($48) per month, at least 60 percent better than elsewhere. Hence, given his level of education, the secondary school graduate is well-off, but not so dramatically as his primary-school leaver counterpart.

In itself a difference in relative advantage might not have a large effect on satisfaction. Studies in the United States have indicated that if a worker is being paid more on his present job than he was on his previous one, he is less likely to be dissatisfied with it. If he is receiving a larger amount, the difference between, say, 20 percent and 30 percent more will increase his satisfaction further only marginally, if at all.[13] Hence the AA with secondary schooling probably does not feel his deprivation relative to the less educated JAA very much; after all, he is better off absolutely. Instead, it is the relationship between the salaries of AAOs and AAs that galls the latter. The AAs now being recruited have the same formal educational qualifications as the AAOs (School Certificate after eleven years) and have received two-thirds as much training in agriculture (two vs. three years). Yet the AAs start with half the salary of the AAOs (see table 5). Thus AAs with secondary education have little difficulty in developing some sense of relative deprivation over the level of their rewards. The result is more prevalent job dissatisfaction among those with secondary schooling than among other members of junior staff in the ministry ($r = -.14$, $N = 169$, $p < .05$).

The negative correlation between education and satisfaction is more than just a product of the ministry's salary structure. All agricultural extension agents do essentially the same work, and March and Simon offer a more general proposition for such a situation: when two workers are confronted with the same job, the one most likely to think it unworthy of his talents is the one with more education (4.6 : -4.12). There are several reasons that this should be true. First, education raises the individual's sense of competence and with it his image of the tasks he is capable of handling. Second, education may well increase a person's sensitivity to his higher order needs. Abraham Maslow has proposed that human beings have a hierarchy of needs. The more their lower order needs are satisfied, the greater their desire for fulfillment of the higher order ones. Thus an individual's first priority must be given to meeting his basic physiological requirements (material needs). After that, he seeks to assure himself of a stable environment (security needs) and then of acceptance and affection from his peers (affiliation needs). The two highest order needs are for self-respect (esteem needs) and self-fulfillment (self-actualization needs).[14]

A person will pursue all of his five needs simultaneously and he is not likely to feel that he has completely satisfied any of them for long. When an individual has completely achieved his initial goals, he quickly evolves still higher targets and expectations. Nonetheless, a man whose wages provide for less of his material needs than he has come to regard as his minimum will usually give this issue priority. Similarly, a person who has achieved his material and security expectations may still press for better pay, but he is likely to attend increasingly to the pursuit of his higher order needs. One of the effects of education may be to raise the individual's expectations of achieving higher order satisfaction (especially esteem and self-actualization). Thus, for example, at the Kenya Bata Shoe Company, Owinoh has found that almost the only employees who express boredom and dissatisfaction with the work on the assembly lines are those with secondary education.[15]

Despite the general validity of the proposition that education stimulates higher order needs, however, it seems to us an inadequate explanation for our findings in the Ministry of Agriculture. Unlike the assembly line, agricultural extension is not intrinsically boring to the educated man. In the United States extension agents are all university graduates. The problems of getting technological innovations adopted are complex and challenging anywhere, and a wide range of potentially satisfying relationships are available with the farmers served. Certainly the Kenyan extension agents themselves are conscious of the interpersonal challenges posed by their work. We asked a subsample of fifty-one staff if they found the work hard and only 27 percent (14) said they did. When asked to explain why they found it difficult or easy, 48 percent (24) mentioned their relationships with farmers. Twenty percent (10) found their farmers troublesome, demanding, or slow; 27 percent (14) felt their jobs were easy because the farmers were responsive and quick to understand. The junior staff can hardly be said to be unaware of the possibilities for higher order need fulfillment in extension work.

Still, the failure of educated agents to respond to the personal challenge of extension work is not unexpected if we follow Maslow's theory. If material and security needs are not adequately met, they take priority over higher order ones. Larger amounts of money make it easier to satisfy these basic needs. By virtue of their wealth, the developed countries have been able to provide relatively well for their workers' basic needs and so their employees' higher order ones have become increasingly important.[16] Almost by definition, the developing countries have not been able to satisfy the material and security needs of their working populations at as high a level.[17] Wages are lower and unemployment is more serious. The basic needs thus have a greater prominence in the developing than in the developed world. The evidence gathered by Harrison and by Kidd in Nigeria indicates that issues of pay and promotions dominate the degree of job satisfaction of extension workers.[18] Of course, African workers are now much better provided for in material welfare and security than they were thirty years ago and so the dominance of these issues is no longer absolute. Our evidence indicates that affiliation needs, the next step in Maslow's hierarchy of needs, are becoming increasingly important to extension agents. The more often a member of junior staff is named a friend by his peers, the greater his interest in a ministry career ($r = .16$; N = 169, $p < .025$). The need for affection is still not as strong as the desire for material benefits, but it is no longer insignificant.

It seems reasonable to assume that the satisfaction of agents also increases as esteem and self-actualization needs are better met, and this effect is probably greater for better educated employees. But the basic issues of pay and advancement overshadow these higher order needs in Africa in a way that they do not in the industrialized world. In a study of a Nairobi firm, Muthaura found that poor supervisory practices, which would have caused workers to quit in the United States, seemed to cause employee dissatisfaction but had no noticeable effect on turnover. The job market was simply too tight for a worker to resign over such a matter. In short, the basic needs for maintenance and security took precedence over the higher order needs for affection and self-realization.[19]

Another way that secondary education might unsuit young men for agricultural extension is suggested by Harrison in his Western Nigeria study. He points out that the bulk of better educated Nigerians come from urban backgrounds and tend to have negative ideas about the rural areas, small farmers, and agriculture in general. Thus the practice of recruiting AAs from among those with good secondary school results automatically leads to employing men with a weak commitment to extension work.[20] In Kenya, on the other hand, very few Africans have been raised in the cities. Secondary school leavers also almost certainly share the feelings of other Kenyans that a man absolutely must have at least a small farm to call his home. Unlike his Nigerian equivalent, an AA assigned to his home area will begin to manage a family farm. Nonetheless, it is true that old extension officers in Kenya complain about the young AAs in their "pointed city shoes." Many recruits will have been in boarding schools for their secondary education, cut off from much direct farming experience and perhaps left with the feeling that the drudgery of agriculture is beneath them. The secondary school leaver does need to be socialized out of his elitest attitudes.

We suspect, however, that secondary education has the largest effects on job satisfaction not because of its influence on attitudes toward agriculture or extension but because of the reactions it creates toward other aspects of work with the ministry. We have already noted the interaction between education and salary expectations. A similar relationship is found with the promotion structure. March and Simon point out that the greater the rate of promotion in the organization has been in the past, the more threatening it is to a worker's self-esteem to have few or no promotion prospects in the present (4.6: –4.13). A rapid pace of promotions at one time creates the expectation that those with similar qualifications will be advanced similarly in the future. When this does not occur, the employee is left to conclude either that he is less skilled than he thought or that his worth is not being recognized and rewarded. This phenomenon presents a major problem to most developing countries, for immediately after independence their citizens were treated to an accelerated pace of promotions that could not possibly be sustained.

Two of the officers we interviewed in Western Province had profited by this Uhuru bonus. At independence they had just completed their secondary schooling and were given a two-year course and made AAs. After a short period on the job, they were taken back for another two-year course and made AAOs. Finally they were sent to the United States and obtained degrees in agriculture. One is now a District Agricultural Officer; the other, a Provincial Farm Management Officer.

Within the space of a few years, they rose from the ministry's second cadre (AA) to its fourth and highest (AO). Now, ten years later, it is almost impossible for a ministry employee to rise from one cadre to another. One must have a given course in agriculture to join a particular cadre and admission to these courses is determined by educational qualifications that the AA and AAO cannot obtain on the job. Worse, although upward mobility is slowing down throughout the civil service, the bars to intercadre movement are not as complete in other ministries. In Education, in particular, teachers can study for the formal examinations externally and so gain admission to the higher level courses. Quite reasonably, Agriculture requires passes in science examinations for admission to its courses, yet one is not permitted to take the higher level science subjects externally in Kenya because of the need for laboratory facilities. Thus today the AA finds himself trapped in his cadre, with less than a third of the maximum potential salary that his equivalents had less than ten years earlier (see table 5).

The object of ambition for the secondary-educated AA is intercadre promotion and, given the slightest opportunity, he will express his concern with it.[21] That it is shut off to him is another cause of his feeling relatively deprived. Of course, intercadre promotion is also more difficult today for the primary-educated agent, but the range of possibilities for those with only primary qualifications was never as great as for those with secondary ones. Furthermore, external studies and upward mobility by that route are not completely closed off to them as they are to secondary graduates in the sciences. Thus the secondary school leavers in the ministry suffer a greater sense of deprivation than do those less educated. This damages their own sense of accomplishment and is another cause of their lower job satisfaction.

We have now examined extensively various influences on job satisfaction, which in turn is the basic determinant of the individual's perception of the desirability of leaving the ministry (4.3:-4.5). At the start of our discussion, however, we noted that the decision to resign from the organization is a function not only of the desire to leave but also of the ease of moving to another job (4.1:-4.3; 4.1:-4.4). We observed that the ease of leaving is basically determined by the number of available job alternatives (4.4: +4.18), although older men are also likely to feel that a job change is difficult even if other employment is open (4.4:-4.21). It is now necessary to return for a detailed examination of the factors that influence the extension agent's perception of the number of alternatives to ministry employment open to him. Of course an individual's awareness of the existence of employment alternatives is partly generated by his propensity to look for them (4.18: +4.32). He who does not seek will not find. Whether or not one searches for other jobs is decided in good part by how satisfying one finds one's present work (4.32:-4.5). Hence the various determinants of job satisfaction have their impact on this part of the resignation decision model. Another factor which weakens the propensity to search is the length of time worked in one's present organization (4.32:-4.33). Time creates habits and weakens resolve to find alternatives even to what seems an unsatisfactory situation. Thus we find the greater an agent's seniority with the ministry, the larger his expectation of continuing to work there ($r = .13$, N = 169, $p < .05$).

The perception of extraorganizational alternatives is almost necessarily dependent on the actual number of jobs available in the economy (4.18: +4.19). The hard reality in Kenya is that jobs in the wage sector are scarce for those without university degrees. Most extension agents are fortunate to have the relatively good jobs that they have secured at the ministry, although they may not all see it that way. Nevertheless, in any economy there are those who will have better access to whatever jobs are available. Individuals who hold high status in a particular society are likely to be more visible to potential employers and so to have better chances for the existing jobs (4.18: +4.30: +4.22). Higher status individuals will also be aware of their relatively privileged position in the job market and consequently regard themselves as having an above average number of alternatives (4.18: +4.22).

The overwhelming determinant of social status in Kenya is education. Hereditary status is virtually nonexistent and wealth presents a weak claim to social esteem compared with scholastic achievement.[22] Furthermore, employers can be counted on to give great weight to education in recruiting new staff, despite its doubtful connection with productivity in a large range of occupations. Education's conferring of social status and its acceptance as a demonstrably objective criterion of selection overwhelm its questionable utility for the particular tasks at hand. As a result, the higher an individual's formal qualifications, the greater his expectations of upward mobility, either through promotions or through interorganizational transfer. In the Ministry of Agriculture, these alternatives and expectations add to the impact of education on job satisfaction to create lower commitment to career employment among those with secondary school qualifications.[23]

To conclude, the degree to which the extension agent is committed to the ministry as his employer is influenced in a very large number of ways by his level of education. It affects his job satisfaction directly and indirectly by several different paths. His education also determines the range of employment alternatives open to him and his perception of them. The Ministry of Agriculture offers a more than competitive package of benefits to its junior staff and few have resigned or are likely to. Nonetheless, the commitment to the ministry of those agents with secondary education is, while positive, relatively weak. Certainly they do not feel any overwhelming debt of gratitude toward the ministry. An appreciation of this fact helps in understanding the negative relationship between secondary schooling and job performance in the ministry.

The Motivation to Produce

The earliest thinkers about employee morale and productivity assumed a stable and uncomplicated causal relationship. They argued that higher job satisfaction would mean the worker took more delight in the work and felt more positively toward his employer, both of which would result in greater productivity.[24] It is now clear that no such simple and consistent causal link exists.[25] Any theory that postulates a simple relationship between morale and output seems to us to be based on three false assumptions. First, it confuses job satisfaction with task satisfaction. If someone finds a task intrinsically pleasing, he will indeed want to perform it more or better. That is what hobbies are all about—tasks so intrinsically satisfying to the individual that he is willing to do them even without any material rewards. Yet a job involves many more things than just the tasks performed: activity, material reward, friendships, and so on. Satisfaction with these other aspects of a job may well overshadow or conflict with the intrinsic pleasure of the task. One could like a job because it demands very little work or because of the opportunities it gives for friendly conversation with other workers. Satisfactions of this sort reduce one's propensity to resign from the job but they actually hamper attempts to increase output on the tasks the job involves.

The second hidden premise in any simple morale-productivity theory is that the worker will credit all satisfactions to the employer. It is obvious, however, that many job benefits are properly attributed to other sources. A general pay rise is satisfying, but it may have been obtained by the trade union, creating stronger obligations to support the union against management. It is pleasing to be accepted on the job by a group of friendly peers, but the social debt that thereby develops is owed to the work group and not the employer and may be used to promote group output restriction. A promotion is a benefit, but if it is attributed to the personal influence of a particular superior, the obligations created will be used for the ends that supervisor personally wills. Other sources of job satisfaction may be credited to God or chance, who are not likely to specify increased productivity as their reward.

The third implicit assumption is that satisfying benefits received from the employer are regarded as creating new social obligations rather than as paying off old ones. This social debt would then produce a desire to perform better in return. Yet the worker may well regard a promotion or an individual salary raise as a reward for past performance or for qualifications already obtained. He is satisfied with his job, for he likes the rewards. But he may regard the benefits as the settlement of an old social debt on the part of his employer. To him they may carry no more obligation to increase his output than does a pension to a retired man.

If the question "How can the job satisfaction of the worker be increased so that he will improve his productivity?" is based on false premises, what then? The generalized concern with morale that it reflects is justified if we are concerned with employee welfare or turnover, but the question put has to be considerably more narrow if we are concerned with output: "How can we provide the worker with satisfactions that he will credit to his employer and that create in his mind social obligations that he can repay through increased productivity?" When the problem is stated in this way, it is not difficult to see why the general level of performance in the Ministry of Agriculture is low. First, employment is likely to be regarded as a reward for good examination results in the past—which creates no new obligations—or perhaps as a consequence of someone's personal influence—which creates obligations to him but not to the ministry. Second, the ministry is quite ineffective in demanding and rewarding good performance, which causes some doubt in the minds of those who do feel an obligation that extension output will fulfill it.

When the job satisfaction problem is formulated as above, it also is easy to see why extension agents with secondary schooling would perform more poorly than their less educated counterparts. Agents with only primary education are far overcompensated by the ministry, even relative to their expectations, and feel some obligation to perform well in return. Within the limits of their capacity, they are willing to do so (though this willingness is often not properly exploited). Because the capacity to do good extension work increases steadily with education, the more primary education an agent has, the higher the level of his performance. The situation for junior staff with secondary education is quite different. First, although objectively they are well paid, they are not as well compensated relative to their education as the primary school leavers are and their pay may not even equal their unrealistic expectations. Second, they are paid far less than the AAOs are. As the latter have only marginally better formal qualifications, this difference generates a great sense of relative deprivation and a grievance against the ministry. Third, the agents with a School Certificate are likely to regard a reasonable salary as a reward for their past examination performances. They probably will not recognize their wage level for what their employer intends—a token of faith in their future good performance, a token that creates a social obligation actually to do well. Given the unrealistic financial and mobility expectations of secondary school leavers in Kenya, it probably is uneconomic for an employer to exceed them and perhaps even to meet them.[26] Does this mean that secondary school graduates will necessarily be undermotivated and perform more poorly than those with less education?

There appear to be two basic ways that more educated extension agents can be motivated to higher standards of work. The first is still to be found within the job satisfaction theories. Implicitly these theories are based on a social exchange model of human behavior, with benefactions, obligations, and output repayment taking place in an atmosphere of mutual trust. As long as we attend only to material benefactions, it is difficult to obtain better productivity through the model, for the expectations of educated agents are hard to exceed and consequently it is very costly to create a sense of social obligation among them. The nonmaterial expectations of the agents are not very high, however, and here it is possible to use social exchange for output purposes. This was the argument made in the preceding chapter. The Kenyan worker expects authoritarian supervision, although he does not like it. This perception of the normal role of authority exists even for university graduates and is certainly strong for secondary school leavers. Education does not seem to alter the individual's expectations of the prevalence of authoritarianism very much, if at all. The teachers he knows are usually demanding, punitive, distant, uncommunicative, and not very helpful.[27] Thus the superior who is willing to be nonauthoritarian is in an exceptionally favorable situation. It is quite easy for him to provide his subordinates with a supervisory relationship that is satisfying beyond their expectations. Those with secondary schooling will feel this treatment to be as exceptional as those at the primary level. The sense of obligation to the superior thereby created can be used by him to raise the organization's productivity (chapter 5). The officer or the LAA who treats his subordinates as semiprofessionals (by being helpful, communicative, and nonpunitive and by allowing them to participate in decisions about their work) seems very likely to be rewarded by them with the kind of job commitment that we anticipate from a professional.

The other basic way that educated agents can be motivated is through the use of incentives. A *morale*-productivity relationship derives from the character of a worker's past and present job satisfaction. The effect of incentives on output, however, is based on the worker's anticipation of future satisfactions. According to the theory of Victor Vroom, the strength of a man's motivation to perform a particular act depends on (1) the balance of satisfactions and displeasures he anticipates deriving from this and alternative acts and (2) the strength of his experience that these outcomes will in fact follow from his act.[28] This formulation is an extension of the Bernard-Simon theory of organizational equilibrium, which we used in analyzing the decision to participate. Vroom's theory, like the Bernard-Simon one, is true by definition but is useful because it helps us to understand, organize, and anticipate a mass of lesser, factually based propositions. Vroom stresses that men do calculate whether the rewards associated with doing something are worth whatever personal costs it may entail. He also points out that this calculation not only involves weighing consequences against one another but also assessing how likely it is that the rewards and costs will in fact follow from his act. Thus a soldier may not consider the prestige that goes with bravery worth dying for. But he may consider the prestige a virtual certainty and his death only a possibility and so be brave anyway.

Although the stimulation of motivation through incentives involves a more calculating, market-exchange type of behavior than does the atmosphere of mutual trust found in social exchange, the two need not be incompatible. Social exchange can take place between supervisors and subordinates while incentives can be provided by the ministerial organization. The most important incentives at the disposal of the ministry are promotions, because they are an outcome from which agents anticipate very great satisfaction. Of course, it is impossible to provide upward mobility to educated agents at the rate experienced during the Africanization period. Thus, it is unlikely that agents will be satisfied with any rate of promotions that can realistically be provided. But their interest in promotions is still such that they will be motivated to produce by the existence of a reasonable number of promotion possibilities if their attainment is tied strictly to performance.

The basic problem is that the ministry has almost completely shut off promotions across cadre boundaries (for example, between JAA and AA ranks or AA and AAO). Because intercadre promotions carry a change of status in addition to one of salary, they are subjectively far more significant to members of staff. The existence of barriers to intercadre movement also limits the number of promotions possible to an individual. The effect of these barriers is illustrated by table 18.

TABLE 18
Commitment to a Ministry Career of Junior Staff
According to Their Salary Grade

| | Commitment | | Number |
	Mean	Variance	of Cases
JAA IV (lowest)	4.00	0.00	3
JAA III	2.74	2.20	23
JAA II	2.97	2.17	60
JAA I	2.33	4.33	3
AA III	3.05	1.76	22
AA II	2.54	2.59	48
AA I (highest)	2.20	3.73	10

$F (6/162) = 1.06; p > .05$
NOTE: Commitment is measured on a five-point scale ranging from 0 (low) to 4 (high).

Although not statistically significant, the differences in the commitment of agents to a career with the ministry show an unusual pattern which can best be explained by the impact of promotion barriers. Promotions within a cadre depend in part upon seniority, so that the higher grades are filled disproportionately by men with longer tenure. We would expect that the more senior agents and those who have been rewarded with the largest number of promotions would have the greatest commitment to a ministry career. Yet as agents get to the top of the JAA cadre and then progress through the AA cadre, commitment to the ministry declines. It would appear that these men are convinced by their promotions that they are talented and feel that because of their diminishing prospects for further promotions, perhaps they should seek employment elsewhere. The barriers to intercadre movement serve as a positive disincentive to these men. This disincentive affects those in the AA cadre, where most of those with secondary education are found, much more strongly than it does those who are JAAs. Movement from the JAA to the AA cadre is difficult but not impossible, where AA access to the AAO rank is now almost unknown. This objective reality is reflected in the perceptions of the agents. In the follow-up interview, 41 percent (14/34) of the JAAs gave themselves a good chance of promotion to the AA cadre. Absolutely no AAs thought they had a reasonable possibility of becoming AAOs and only 18 percent (3/17) gave themselves any chance at all. Thus the disincentive of intercadre barriers to promotions hits the AAs and those with secondary education the hardest.

There are a variety of reasons why intercadre promotions should not be barred but rather used as a positive incentive. First, given the geographic dispersion of extension work and the collective organization of the junior staff, the agents are simply too powerful to be forced to produce more. To be effective, any efforts to increase the output of extension workers will need their willing cooperation. Second, inter-cadre promotions are the most powerful incentive it is possible to offer to the junior staff. Third, the main reason for having barriers to intercadre movement in the first place was that more education was presumed to be necessary to good performance at the higher levels. Since we have shown that education is more often a hindrance than a help to good extension work at present, the rationale supporting the continuation of the barriers is invalid.

An incentive-oriented promotion system could be constructed in a number of different ways. By way of example, we give here the set of reforms that we proposed to the ministry in an earlier set of organizational recommendations:

1. Permitting no one to enter directly one cadre who would not be capable of eventual upgrading to the next. If, for example, a grasp of science is essential to the performance of an AAO's duties, those who get the agricultural certificate and become AAs should either have that competence when they enter or get it at the training institution.
2. Selecting each year a substantial portion of those with 10 to 15 years' service in one cadre for upgrading to the next. The object would be to offer each man entering the Ministry a 25 per cent or

better chance of being upgraded to the next cadre midway in his career. This would provide him with a significant incentive, the effect of which would be felt throughout the 10 or so years in which he was striving to qualify. If the 25 per cent formula were followed, half of each cadre would be made up of upgraded men and the other half of upgradable ones.[29]

3. Basing promotions and upgrading overwhelmingly on job performance and not on potential, formal qualifications, or seniority. Point 1 would guarantee that all members of a cadre had the minimum potential for going up a rank. Further distinctions on the basis of potential would undercut the object of the restructuring, which is to increase job performance. Included in the concept of job performance, however, could be scores on agricultural information tests which members of a cadre took regularly throughout their careers in order to prod them to keep informed. Nonetheless, the greater part of performance would be judged from their superior's periodic objective evaluations of their field work (for AAs and JAAs of their demonstrations).

4. Leaving some promotion opportunities within each cadre for attainment after the 15 years of service mark. In this way men who are passed over for upgrading would still have some incentives for working.

5. Providing training courses for the men being upgraded to a cadre that are shorter than, somewhat different from, and largely separate from those provided for new entrants to a cadre. The current practice is to put those few men selected for upgrading through the same programme as young men fresh out of secondary school. . . . This is wasteful. The staff members being upgraded already have some training in agriculture and have considerable field experience. Further, they are likely to have their greatest difficulty in the more academic areas. Trainees fresh out of secondary school, however, will be at ease in the bookish courses and will need a thorough introduction to the practicalities of farming and extension work. When staff in for upgrading are thrown in with new recruits, their time is largely wasted in the practicals and they are over their heads in the academic portions. (A similar conclusion is suggested for Western Nigeria. . . .)[30] Courses for experienced men need to take their background as a starting point, avoid repeating it, and give extra attention to the theoretical courses, where they will have problems. Even though the training programmes for new recruits and upgraded staff need to be largely different, there are areas of overlap which would justify

keeping them in the same institution. Furthermore, the older staff may be able to indoctrinate informally the new recruits into the realities of extension work.[31]

We do have some direct evidence that such a set of reforms in the promotions incentive structure would in fact raise the performance level of the secondary School Certificate holders. As is illustrated in table 19, junior veterinary staff do show a uniform improvement in their performance as their education increases. This dramatic divergence from the situation with agricultural extension agents seems to result in large part from the veterinary agents' perceiving their promotion possibilities

TABLE 19
Total Information Scores on Feeding Grade Cattle
According to the Education of Junior Veterinary Staff

Education	Mean Score	Variance	Number of Cases
Standards I–V (1–4 years)	6.9	10.12	8
Between Std. IV and C.P.E (5–6 years)	7.9	9.78	15
Certificate of Primary Education (7 years)	9.8	23.30	12
Kenya Junior Secondary Examination (9 years)	10.5	40.50	2
School Certificate (11 years)	12.7	13.24	7

$F(4/39) = 2.68; p < .05$

differently. The School Certificate holders who went to the Embu Institute of Agriculture for training were told quite truthfully during their course that they should expect to spend their careers within the AA cadre. Those who were sent to the Animal Health and Industry Training Institute, however, were given quite another set of expectations. The ambitious instructors at AHITI believed that their certificate graduates were better qualified in veterinary medicine than the diploma graduates from Egerton College and told their students so. They led their Animal Health Assistants (AHAs) to expect that those who did well in the field would be called back in a few years for upgrading to the next cadre as Livestock Officers (LOs).[32] Unless a change in ministry policy occurs, the expectations of promotion that AHITI fostered were actually false and eventually could lead to bitter disappointment. At the time we did our research, however, School Certificate AHAs believed that major promotions were available to those who performed well in their jobs. The effect of intercadre promotion incentives upon extension agent performance thus can be seen in table 19 to be substantial enough to more than overcome the negative effects of secondary schooling on motivation.

The strong incentive effect of promotions is not unique to Kenya. The American research on this subject suggests that advancement prospects constitute the first or second most powerful factor motivating employee behavior. Furthermore, there is some evidence that promotion possibilities become increasingly important to workers as their education increases.[33] Promotions offer many different forms of satisfaction. Not only do they bring increases in pay but also demonstrate the high regard in which one is held by others and help to meet one's desires for self-fulfillment. In Maslow's hierarchy of human needs, promotions simultaneously satisfy both basic and higher order needs. Thus the effectiveness of advancement prospects as an incentive is probably independent of the general level of need satisfaction in a society and is therefore a genuinely universal aspect of human behavior in utilitarian organizations.

The Determinants of Promotion

In setting out Vroom's theory of motivation, we noted his postulate that the strength of an incentive is dependent on how likely the worker feels it is that the outcome he desires will in fact follow from the action the incentive is supposed to encourage.[34] If promotions in the Ministry of Agriculture are designed to encourage good extension work, how probable is it in the agent's experience that good work will produce a promotion? If the actual relationship between the two is very small, promotions could hardly be expected to have much effect on productivity. Lawler has argued recently that one of the reasons pay has appeared in many studies to be a relatively unimportant incentive is that it often is not related to merit on the job.[35] We have already noted that none of the AAs and only 41 percent of the JAAs believe they have a reasonable prospect for intercadre mobility. In addition, 38 percent of the junior staff feel that their prospects for promotions of any kind are poor. But what do they perceive to be the factors determining the promotions that are made?

In our follow-up interview, we presented to each agent a list of twelve items that we said might affect his chances of promotion in the ministry.

We then asked him to rate how important each one was for promotions and followed this by requesting that he single out the first, second, third, and fourth items in significance. The pattern of responses to these questions is presented in table 20. It is striking that formal agricultural

TABLE 20
The Perceptions of Agricultural Extension Agents of Factors in Promotions

Approximate Order of Importance[a]	Factor	Percentage Terming It First in Importance	Weighted Score of Importance[b]	Percentage Terming It Very Important
1.	The amount of training you've had in agriculture	27	117	92
2.	The amount of your formal education	27	88	77
3.	How hard you work	4	82	94
4.	Your technical knowledge in agriculture	12	80	98
5.	Your obedience to your superiors	15	65	90
6.	The extent to which you show initiative and do things without being told	8	29	69
7.	Your intelligence	4	19	65
8.	Whether you know someone personally who had the influence to get you a promotion	4	18	29
9.	How well you keep your own farm	0	10	50
10.	Your ability in English	0	9	15
11.	Your tribe	0	3	4
12.	Your physical fitness[c]	0	0	2

[a] This order is based on the Weighted Score of Importance. It is not the order in which the factors were read to the respondents.

[b] This score is calculated by awarding 4 points to a factor each time it is named first in importance, 3 points for second, 2 points for third, 1 point for fourth, and then adding the points.

[c] This factor has no relationship to extension performance and was specifically included as a validity check. As can be seen from the responses, only one agent termed it important for promotions.

and educational qualifications are ranked as the two most important factors in determining one's promotion prospects, because these are attributes totally independent of one's performance on the job. In fact, we find that 29 percent of the agents (15/52) credit overwhelmingly predominant influence to characteristics that are determined before they ever began work (items 1, 2, 7, 10). A further 38 percent (20/52) consider these formal sorts of criteria to be equal in importance to all other factors influencing promotion prospects.[36] Fully 58 percent of the junior staff think that one or the other of these four items is the single most important determinant of promotion. This belief in the signifi-cance of formal qualifications almost certainly weakens the effect of the promotions incentive on job performance. If one is a JAA with a secondary School Certificate (eleven years) and one thinks that this qualification in itself makes it likely that one will be taken for upgrading to an AA, there seems little reason really to exert oneself in one's work. If, on the other hand, one is an AA without a Higher School Certificate (thirteen years) and one is convinced that access to AAO training is dependent upon that qualification, there also is no incentive to do a good job. Of course, very few agents believe that work performance is insignificant in deciding one's promotion chances, so some incentive effect does remain. The effect would be much stronger, however, if the importance of formal criteria were downgraded.

Of course there are ways other than the use of formal qualifications in which nonperformance criteria can influence promotions and distort their incentive effect. Jon Moris makes a strong argument to this effect:

> No independent, professional criteria of performance are recognized; nobody asks the farmers how they feel about a particular service or official. Instead, the supervisor's annual confidential report is still the main basis for evaluating an officer's field performance. As a consequence, within the bureaucracy achievement is noted mostly in default—actions do not become individually identified unless they are mistakes. . . . Junior officers see that while their procedural derelictions or even differences in opinion (which may be profes-sionally based) are interpreted *personally* and are remembered, positive achievements are not. . . . The junior officers jockey endlessly with each other to keep alive their direct, personal ties to influential men above. As perceived by them from the inside, what matters is not what one does, but rather whom one knows.[37]

As we will stress in chapters 9 and 10, junior staff performance is judged subjectively, imperfectly, and mainly in default—all of which weakens the incentive effect of promotions. Moris's emphasis on personal subservience and influential contacts, however, is misplaced for the extension agents (although it is almost certainly appropriate at the senior staff levels). The AAs and JAAs give little importance to personal contacts in gaining promotions (see table 20, item 8.) They see the much-discussed factor of tribe as virtually insignificant. (They probably believe tribalism exists but find it has no meaning for them as their competitors for promotion are in their province and belong to the same tribe.) It is true that 42 percent of the junior staff see personal subservience to a superior as one of the four basic determinants of promotion (items 5 and 8). Yet the other 58 percent stress the importance of work performance itself (items 3, 4, 9) and 19 percent consider independent initiative to be among the four attributes most contributing to promotion. The assessment of junior staff performance is imperfect and subjective but it is not as personalized or politicized as Moris would suggest. There is much room for improvement here (as we will suggest in chapter 10), but promotions do still offer some incentive to real extension performance. It is precisely for this reason that the differential availability of intercadre promotions to JAAs and AAs depresses the relative performance motivation of the latter.

Summary

To retrace our argument, secondary education has a uniformly detrimental effect on the work performance of agricultural extension agents in the current Kenyan context, one similar to that of many developing countries. Nonetheless, we can see no strong reason why further schooling should harm one's capacity for extension; in fact, it should improve it. Thus we were led to examine the impact of education on employee motivation. In the first instance, we analyzed how secondary education in Kenya lowers the level of an agent's job satisfaction and commitment to an extension career, for it creates employee expectations that can no longer be fulfilled. Because secondary school leavers are less satisfied with their work than primary educated men are, they feel relatively less obligation to the ministry to do their work well. This problem of disappointed expectations and consequent relatively low job satisfaction cannot be dealt with practically speaking through increased material rewards, for the levels of expectation are too high. It is probable, however, that the job satisfaction of secondary educated agents could be improved through treating them to a less authoritarian and more professional set of supervisory relationships. Nonetheless, the most effective way to deal with the lower performance of secondary school leavers is found by leaving the satisfaction-motivation model and turning instead to an incentive one. The example of School Certificate veterinary agents (AHAs) demonstrates that they can perform better than primary leavers when they believe that good work will lead to substantial promotions. Of course, for promotions to have their maximum incentive impact on performance, it is essential that agents believe that upgrading is indeed primarily a function of the quality of their work. At the moment, too much reliance is placed upon formal qualifications and (to a lesser extent) personal subservience for promotions to have their maximum incentive effect.

The foregoing leads us to conclude that the educated worker presents a delicate motivational problem in current Kenyan working conditions. The employer who is not prepared to find solutions to these problems would be better off with employees of lesser formal qualifications. Yet incentives do exist, particularly promotions, that are able to counteract the negative influence of the educated workers' high expectations. When properly used, these incentives can unlock the educated employee's capacity for better work and amply repay his organization's investment in him.

The fundamental point is that the problems of administration in developing countries cannot be overcome by a simple, frontal assault on the training of personnel. Higher levels of skill create new incentive problems which themselves have to be solved if the new capacity is to be used fruitfully. The issue is well stated by Norman Uphoff:

> In evaluating bureaucratic failings, in Ghana and other African countries, I would focus on four factors: (1) competence, (2) organization, (3) incentives, and (4) the environment. Some would focus on one or another even to the exclusion of the others, but I see them as mutually reinforcing. This is to say, in economic language, that there are rapidly diminishing returns to improvements made in any one area without complementary supporting changes in others. For example, as many donors have discovered, training (to increase competence) without more appropriate organization and more compelling incentives for better performance, will yield few benefits.[38]

The educational levels of administrators need to be raised in conjunction with improvements in the incentive systems governing their work, so that they will be motivated to make good use of their new competence.

In making our analysis of overeducation, we were able to use a variety of American motivational theorists, all to good effect. The social exchange model of behavior, and the theories of March and Simon, Abraham Maslow, and Victor Vroom all provided pieces to the puzzle of the poor productivity of the secondary educated extension agent. This is not to say that Kenyan workers are motivated exactly as American ones are; the contrary was found to be the case. But we did find that the United States theorists have built models of motivational behavior capable of encompassing and explaining the basic Kenyan variants.

7 Communications:
Teaching the Teachers

The Ministry's Information Transmittal System

The junior agricultural extension agent is the vital link in the chain of communications between the agricultural research station and the small farmer. Very little information on technological innovations reaches the small farmer without passing through an AA or JAA. New farming practices are rarely learned from formal schools, commercial contacts, or the mass media. Similarly, the direct contact of small farmers with senior agricultural staff and research stations is negligible. The only significant source of information that is genuinely independent of the junior extension agents is work experience on a commercial farm. Even this source accounts for only a small portion of the new information (between 6 and 20 percent) and application of the new practice thus learned is delayed until the worker's retirement or extended vacation.[1]

As the small farmer has very little access to information on innovations other than through his extension agents, the communication processes by which these junior staff are kept informed are of great importance. There are many possibilities for breaks and distortions in the chain linking the farmer to the sources of technological information. Experience indicates that these possibilities are all too often realities in the poor countries. Yet our knowledge of the communication processes within the extension services is impressionistic only. It is critically important that we systematically investigate and analyze the ways that the teachers are taught so that the instruction can be improved.

The extension agent obtains his technological information through a considerable variety of communication links, as Figure 4 illustrates diagrammatically. The basic and applied research leading to the recommendation of agricultural innovations is done in a variety of international research bodies, the various Kenyan research stations, and the universities. The Kenya government and the several crop boards have built up an impressive complex of agricultural research stations, which are doing quality work. Their findings become a basic part of the curriculum in all the institutions engaged in training agricultural

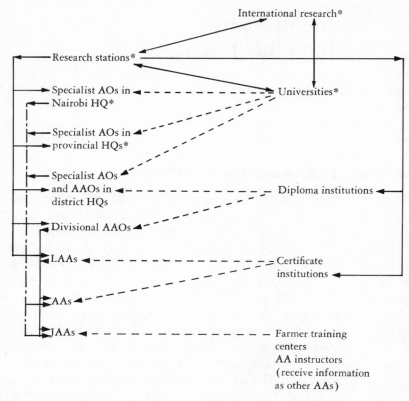

Key:

———— Relatively routine channel of communication (used 3 + times per year)
.—.—. Refresher courses
-------- Initial training courses
   *   Points at which technological information is created or significantly adapted

Fig. 4. Channels by which junior agricultural extension staff receive technical information.

officers and extension agents. In addition, the research stations do a fairly good job of maintaining regular direct contact with all levels of senior staff who are doing work related to their particular specialties. Mimeographed research reports are issued to Agricultural Officers; at least some national speciality stations arrange annual tours in which they hold provincial conferences with AOs and AAOs, inform them of recent developments, and seek information about field problems that have been encountered. Local, general stations hold annual field days for the area's agricultural staff. Our impression is that the Kenyan research stations maintain a reasonable concern for the communication and utilization of their findings by all levels of ministry senior staff.

Kenya's research stations do much of the development as well as the research on their agricultural innovations. Nonetheless, a few of the Ministry of Agriculture technical divisions also do their own work in refining and adapting research findings for regular farm use. In these specialty areas, regular Agricultural Officers at the national and even sometimes the provincial level are engaged in developing packages of farm husbandry recommendations. The officers in these technical divisions are not only transmitters of technical information but creators and adaptors of it as well.

Two patterns can be found in the ways that various technical divisions communicate information to their field staff. In one, the national and provincial officers are active and supplementary transmitters of information between the research stations and field staff. These men come to be regarded as experts in their own right on their specialty, and field officers with a problem will usually consult them rather than the relevant research station. In the other pattern, the national and provincial officers in a technical division play a primarily administrative role and leave the communication of information and technical problem-solving to the research stations. Our impression is that at the time we did our research, these patterns were rational and complementary. Technical areas in which the research stations were uncommunicative had officers in the ministry headquarters who played the role of expert and information disseminator. Those specialties for which the national officers had assumed an administrative role had vigorous and outgoing research stations. Whether these patterns were permanent and which side of these complementary relationships was cause and which effect, we did not investigate. We did sense, however, that good quality technical information was being made available to at least the Agricultural Officers who wanted it at all levels.

The information transmittal system we have just outlined reaches only the senior staff in the ministry. The communication network for junior staff is basically distinct and different. Of course, extension agents attend separate training institutions from the senior staff: AAs gain their certificates at special Institutes of Agriculture; JAAs are given short-courses at local Farmer Training Centres. But in addition, junior staff have no routine contact with the research stations and national and provincial specialists who create the recommendations they are expected to disseminate. They neither read their reports nor meet with them informally or around conference tables. Direct contact of AAs and JAAs with the sources of their information is confined to occasional refresher courses organized on a provincial basis, where provincial specialists and sometimes national and research specialists may lecture. Unlike senior staff, the extension agents have almost no opportunities to gain their knowledge of technical innovations firsthand. Instead, information generally reaches junior staff through one or more hierarchical filters. The most direct routine channels of communication to the AAs and JAAs are refresher courses or conferences. These may be held in the open on a payday or organized into a residential seminar at a local Farmer Training Centre. Most often these are addressed by divisional and specialist AAOs and district AOs. Less often provincial AOs give the lectures. The other routine communication channels are more strictly hierarchical: district specialists, who have received their information from their superiors, talk informally to the junior staff in their speciality; divisional AAOs, who have been provided with standard recommendations by the district, pass them on in staff meetings; AAs in charge of locations (LAAs) give talks to their AAs and JAAs at the junior staff gatherings which sometimes precede the weekly chief's *baraza* (public meeting). We turn now to an analysis of the operation and relative effectiveness of these various junior staff information channels.

An Information Transmission Model

The social science literature is strangely deficient in analysis of the kind of hierarchical, downward communication of information found in an extension organization. Sociologists of organization have given considerable attention to upward information flows; experimenters have produced a good deal of data about communications in small groups; educational theorists have written endlessly about information transmittals between teacher and student; communication specialists have done much research on the effectiveness of various media for diffusing innovations; but the literature shows relatively little evidence of either research or theoretical analysis specifically on the problems of passing information down the multiple hierarchical levels of a large organization.[2] We would like to begin to fill this gap by drawing upon the communications literature to provide a theoretical model for this aspect of organizational behavior. We can then analyze our data on Kenyan extension using this model. The conclusions that we draw can then be used in understanding and correcting communications problems in similar hierarchical organizations in other poor countries.

We will begin by specifying the components of our model in a two-person communication system and then extending them to a multiple hierarchy system. Our two-person model is a modification of that described by David Berlo in *The Process of Communication*.[3] The first components are the communicators themselves: a sender and a receiver. The one has a message that is supposed to be communicated for some purpose to the other. A message can vary in its complexity, and the more complex it is, the more difficult will be the acts of transmitting and receiving it. The sender will have a certain motivation to transmit his message and the receiver, a particular motivation to receive it. The motivation to send or receive a message also can vary and essentially determines the amount of energy that the communicator is willing to expend in the act of communicating. In order to transmit his message, the sender will have to put it into some sort of communicable form; he will have to speak it or write it or make another type of signal. When he does so, he is using an encoder. It is useful to think of the sender as having several different types of encoders. Different communication channels require different types of encoding; the communicator can encode his message into a conversation, a lecture, a letter, a pamphlet, and so forth. The codes used also can vary considerably in their complexity: a discussion of the concept of life can be encoded in the simple language of a child or the highly technical jargon of a bio-chemist. Furthermore, encoders differ in the fidelity with which they transmit the message. Variable proportions of the original information will be lost or incorrectly transmitted by different encoders. Finally, the use of one encoder may require considerably more energy or effort on the part of the sender than another one. The ratio between the encoder's fidelity and its use of energy can be considered its efficiency.

The message has now been sent and enters a particular communication channel. This channel, like the encoder, has certain fidelity characteristics. Some portion of the message will be lost or distorted in the channel. At the other end of the channel, the message is received by a decoder. The particular decoder used by the receiver depends on the character of the message sent. The channel and complexity of the message are set. Assuming that the receiver has a decoder for a particular channel (that is, he is able to hear, see, or read), what is the capacity of his decoder for message complexity? Like the encoder, the decoder also will have an efficiency character, that is, it will work at a certain level of fidelity and require a particular amount of energy. The amount of energy the receiver is willing to expend will depend on his motivation. The end product is then a message received, which will contain a certain proportion of the original information transmitted, the rest having been lost or subverted through distortion.

Let us illustrate this model by comparing and analyzing the hypothetical communication of some technical information on grade cattle between a Ministry of Agriculture Research Officer (RO) and a district AAO specializing in animal husbandry on the one hand and between the same RO and a JAA on the other. The RO had developed a certain set of findings and recommendations on feeding grade milk cows. Since the optimal feeding pattern varies according to the cow's milk production and stage in the calving cycle, the message to be transmitted is fairly complex. The RO's job and his professional self-esteem depend on the recognition he receives from his colleagues for his recommendations, so he is highly motivated to communicate them in some form. He chooses to write a paper, for this is the standard channel for scholarly communication. He also decides to write in the fairly complex code of technical jargon. He does so because he is most efficient in this code (he can encode with the greatest fidelity for the least relative expenditure of energy). He works on the paper reasonably hard and very little relevant information is lost or distorted; the encoder had high fidelity. Let us assume that the paper was put onto stencil by a good typist and run on a reasonable mimeograph machine. Thus the fidelity of the communication channel was very high and almost no information was lost there.

This paper is then received by the AAO. The channel decoder is already set for him—reading. The AAO has been reasonably well trained, so he is capable of handling a message of this technical complexity. But he is by no means as efficient as the RO at this complexity level and in order to decode with a reasonable level of fidelity he will have to expend a fair amount of energy on the decoding. The paper is in the AAO's specialty, so he is motivated to read it. He has no direct, personal contact with farmers, however, and he has been passed over for promotion twice. Thus he has no need to impress either his clients or his superiors and will not expend too much energy on the reading. As a result, he reads the paper through once carefully but he does not reread and try to understand the particularly difficult passages. The end product is that the AAO has received a message which represents 65 percent of the information the RO had to transmit. A diagrammatic representation of this process is given in the first example of figure 5.

Communication 1: RO MS $/-$ (C F E $>$ ——— $<$ C F E) $-/$ MR AAO
                        $+ \ + \ + + \ +$ paper $= = = =$ 65

Communication 2: RO MS $/-$ (C F E $>$ ——— $<$ C F E) $-/$ MR JAA
                        $+ \ - \ = = \ -$ lecture $- - - \ +$ 45

Key:

The level of a particular component is given below its symbol.

RO	Sender (Research Officer)
MS	Message sent, complexity
/–	Motivation to send
(>	Encoder
C	Complexity
F	Fidelity
E	Efficiency
———	Channel
<)	Decoder

–/	Motivation to receive
MR	Message received, percentage of original information
AAO JAA }	Receivers
+	High amount of
=	Medium amount of
–	Low amount of

Fig. 5. A diagrammatic representation of two hypothetical simple acts of communication.

Now let us examine the communication process between the same RO and a JAA. Professional papers are clearly inappropriate for communicating information to junior staff, so the Nairobi head of the Animal Husbandry Division persuades the RO to address a group of AAs and JAAs who work in areas with many milk cows. There is very little professional payoff to the RO for this kind of exercise, and he is not motivated to spend much energy on preparation. He does recognize that an audience of AAs and JAAs will not be able to handle much complexity and so when he speaks he encodes his message in nontechnical language. He is not very efficient at this level of complexity, however, and as he did not prepare much he leaves out some of the information and slightly distorts other parts. The communication channel, a lecture in a large hall, also lacks high fidelity and certain parts of the message get lost through noise and distractions. Our JAA had had very little training in agriculture and has finished only primary school. He therefore has a very low capacity for complexity both in animal husbandry and in English. The result is that he will have to struggle very hard if he is to understand what is being said. His decoder for this type of message and channel is quite inefficient and is incapable of perfect fidelity in any case. Nonetheless, the JAA is highly motivated: he has many farmer-clients who are asking for precisely this information and he still hopes to impress his superiors and gain access to an upgrading course and AA status. The result is that he puts a great deal of effort into his decoding of the lecture and despite his low capacity and efficiency is able to receive 45 percent of the message that the RO originally had to transmit. The comparison between these two communication processes is presented diagramatically in Figure 5.

We need to add one component to this basic model in order to use it in the analysis of multiple-stage communication processes—the element of the delay between the time an intermediate communicator receives a message and the time he sends it on. The longer the delay, the larger the proportion of the message that will have been lost through imperfect memory. (It is crucial to note here that if one rereads a paper or looks through one's notes on a talk one is receiving a second message and delay is thereby avoided.)

We can illustrate the multiple-stage communication model and the importance of delay in it through the following example, which is based on actual figures derived from our field interviews.[4] Figure 6 presents a diagrammatic analysis of the process of communicating information on hybrid corn fertilizers from the research station to an Institute of Agriculture, on to an AA and finally to a farmer. Information transmission begins with the Research Officer, who has a package of recommendations of only moderate complexity which he would like to have adopted by Kenya's small farmers. He writes a paper on the subject and has it distributed widely. This paper is read by the lecturer on crop husbandry at the Institute of Agriculture. As he teaches this material to future AAs, he is highly motivated to learn it himself. In addition, he has been well trained and has no difficulty reading a technical paper. Because the sender and receiver are highly motivated to communicate and because the one's encoder is well matched to the other's decoder, the communication is a particularly successful one, with something like 94 percent of the original information being successfully received.

There is some delay before the lecturer actually teaches this material

RO MS /— (C F E > ———— < C F E) —/ MR IoA * MS /— (C F E >
      = + + + +   paper   + + + + 94   —1 88 + = + +

———— < C F E) —/ MR AA * MS /— (C F E > ————
lecture   = + + + 80   3 41 + — = +   visit

< C F E) —/ MR AAF
      — = = = 29

Key:

*	Delay in years
AAF	Above-average farmer
IoA	Institute of Agriculture

Other symbols as in figure 5.

NOTE: All figures given under the symbols MR and MS represent percentages of the total amount of original information first communicated by the RO.

Fig. 6. A diagrammatic representation of a multiple-stage communication process in the Ministry of Agriculture.

and he forgets some of it. But the time lapse is well below a year and he is still able to transmit 88 percent of the full amount of information (or 94 percent of what he received). He is a good teacher and is interested in getting his material across. Assuming that his students have finished primary school but have not earned any secondary school qualifications, they are well motivated and able to understand what the lecturer has to say. Once again we have strong motivation in both the sender and the receivers and the encoder and decoders are well matched. The result is that these future AAs receive 80 percent of the original information (or 91 percent of what was taught). We might note that this 80 percent is almost certainly higher than would have been obtained in a direct communication between the RO and the AAs. The RO's encoder would have been too complex for the AAs' decoders, and he would not have been particularly motivated to make the communication effective. In this case, the two-stage communication process is definitely superior in all respects to the single-stage one.

Our AA leaves the institute and goes into the field to work. Let us assume that three years pass without his receiving any additional communications at all on hybrid corn fertilizers. (Such a lapse is actually fairly uncommon in the ministry, but the assumption will help us to illustrate the effect of delay.) At the end of this period he wishes to pass on the fertilizer recommendations to a farmer. The result of this delay between receipt and onward transmission of the message is a substantial loss of information—opposed to the 80 percent he knew at the institute, the AA now knows only 41 percent (thus having forgotten 49 percent of what he had learned). He is fairly well motivated to pass on this remaining information, and he is able to encode the message at a fairly low level of complexity, matching that of the farmer's decoding capacity. The AA is efficient at this sort of communication, although the message does lose a bit in the process of being simplified and passed on. The farmer to whom the AA is communicating is an above-average one: he is able to grasp the message in its simplified form but not without some loss of information. He also does not grasp new farming concepts quickly and so he needs to be highly motivated in order to give his full attention. Moreover, his interest in this new practice is only moderate. The result is that he receives 71 percent of what the AA says, which is only about 29 percent of the information that first left the research station four years before. This is a poor proportion and we will discuss ways of increasing it. Nonetheless, it is worth noting that this is more information than the farmer would have received had the AA not existed. Neither the RO nor the institute lecturer would have had the time or motivation to make such a farm visit themselves. The delegation of information is just as important to organizational efficiency as is the delegation of authority.[5]

There are several general points that can be drawn from this example which will be useful to us as we think of the totality of the information transmission system in the ministry. First, each act of communication entails some loss of information. When all other things are equal, an increase in the number of links in a communications chain leads to the loss of more information. Second, some receivers have decoders that are poorly matched to the encoders of some senders. These receivers will actually obtain more information if an intermediate transmitter is added that has a decoder well matched to the original sender's encoder and an encoder compatible with the decoder of the final receiver. In this type of case, adding an extra link to the communications chain actually improves the effectiveness of message transmission, for here, contrary to the assumption made in our first point, other things are *not* equal. The matching of encoder and decoder characteristics is important to high fidelity communication. Third, good motivation to communicate is necessary in both the sender and the receiver if a large proportion of the information is to be transferred. It is worth noting that motivation varies from time to time for individual communicators. This motivation is often greatly influenced by the position of the other person in the communication dyad. The RO was more concerned about passing information to his fellow professionals than to an audience of junior staff. Similarly, an AA will pay more attention to a message from an institute lecturer than to one from a fellow AA. Fourth, in multiple-stage human communication the delay between acts of receiving and transmitting information is a major determinant of the quality of the system. Long delays lead to substantial information losses. Basic to all the above four points is that an intermediate transmitter cannot pass on more information than he himself has received and retained. Fifth, some delegation of information transmitting is essential in an organization the size of the Ministry of Agriculture. The problem is to identify the pattern at which the optimal amount of information is actually being received. This concerns the efficiency of the total information system. With our model and these points in mind, we can now pass to a detailed analysis of our actual data.

The Factors Affecting the Staff Learning Process

Methodology

Junior staff receive their agricultural information through several different channels, as we noted in our discussion of figure 4. As the influences of these several channels overlap, we need a fairly sophisticated method of analyzing our data if we are to identify the relative importance of each one separately. The techniques used here are those of multiple regression analysis. As explained in chapter 2, the dependent variables in our analysis are the tests of Agricultural Informedness and of Explanatory Ability which we gave on hybrid corn fertilizers, the feeding of grade milk cows, and pest control in coffee or cotton. All staff were tested on the first two topics and were questioned on either coffee or cotton, depending on the area in which they were working. The coffee and cotton scores were adjusted to have the same means and standard deviations and then were combined, giving us a single measure on market crop pest control for all junior staff. As a result we have six separate dependent variables—three tests of Agricultural Informedness and three measures of Explanatory Ability.

In table 21 we present the zero-order correlation coefficients between the six dependent variables, on one hand, and the various independent

TABLE 21
The Correlates of Agricultural Informedness and Explanatory Ability in Junior Agricultural Staff

	Agricultural Informedness			Explanatory Ability		
	Corn Fertilizers	Grade Cattle Feeding	Pest Control in Coffee/Cotton	Corn Fertilizers	Grade Cattle Feeding	Pest Control in Coffee/Cotton
A. *General Training*						
1a. Education: Upper Primary	.23	.13	.17	.14	.11[a]	.13
1b. Education: Secondary	-.11[a]	-.07[a]	-.13[a]	.01[a]	.04[a]	-.08[a]
2. Fluency in Swahili	.09[a]	.14	.20	.16	.16	.12[a]
B. *Agricultural Training*						
3. Holds Agricultural Certificate	.17	.33	.20	.26	.32	.13
4. Was taught with demonstration on topic	.13	.33	.21	.19	.28	.36
C. *Rebriefing on Topic*						
5a. Within last year	.30	.33	.30	.20	.25	.46
5b. By senior staff member	.34	.37	.34	.28	.29	.46
5c. At FTC	.19	.36	.36	.23	.21	.25
5d. With a demonstration	.26	.44	.34	.26	.41	.43
5e. Combined index of a–d	.36	.44	.42	.31	.34	.49
D. *Reinforcement*						
6. Specialized in topic	…	…	…	…	.11[a]	.20
7a. Informedness of AAO on topic	…	.36	.26	…	.16	…
7b. Explanatory ability of AAO on topic	.26	…	.29	.09[a]	…	.04[a]
8. Number of location staff meetings	.15	.28	.32	…	…	…

9a. Informedness of LAA on topic	.38	.41	.16	.29	.21
9b. Explanatory ability of LAA on topic	.35	...	...	.28	...

NOTE: All figures are zero-order correlation coefficients. N = 169. All coefficients of .127 and above are statistically significant at the .05 level for a one-tailed test.

aThese coefficients are not statistically significant. In the case of 1b they are included to demonstrate that performance does in fact fall with secondary education. In the other cases they are included because further analysis leads us to assume that a positive relationship exists even though it may not be statistically significant in the particular case. Where no coefficient is reported the relationship is not statistically significant under any conditions.

variables that influence them on the other. These coefficients represent the simple relationships between the variables and ignore the possibility that several closely related independent variables actually may be combining to produce the apparent effect on the dependent variables. This overlapping of effects can be removed through multiple regression analysis and the results of such an exercise are given in table 22. Here the only independent variables shown as related to a particular dependent variable are those that can be demonstrated to have an impact distinct from and additional to that of the other independent variables listed. From table 22 we can conclude just what are the various independent influences on junior staff informedness.

The final step in our analysis is presented in table 23. From table 22 we are able to establish the various basic determinants of junior staff informedness. We then proceed on the assumption that determinants that are important in some technical areas usually will be influential in all. Thus we give a multiple regression equation for each dependent variable that uses all the independent variables. The regression coefficients from these equations measure the impact of each influence in each technical area. These measures in turn can be used to estimate the average influence of that factor on Agricultural Informedness or Explanatory Ability. Although both tables 22 and 23 are based on multiple regression equations, they are different because they serve dissimilar purposes. The first is designed to show what causal factors can be said with statistical significance to be at work. Table 23, on the other hand, measures the amount of impact of the causal factors that have been demonstrated to exist in table 22. The question of statistical significance is vital to table 22 but is irrelevant to table 23.[6] Having established these methodological points, we can discuss the substantive meaning of the data themselves without interruption.

Decoding Ability and Receiving Motivation

We can begin with an analysis of the factors that affect the decoding ability and receiving motivation of the junior staff. The decoding ability of a receiver is determined by his capacity to handle complexity in the messages and his efficiency in dealing with particular message channels. Since most technical messages are communicated to junior staff orally in the ministry, we would expect that the agents' decoding ability would be improved by the various forms of education that increase their depth and efficiency with language (both in general and in agriculture). An excellent example of this concerns the capacity of the agent to under-

TABLE 22
The Controlled Correlates of Agricultural Informedness and Explanatory Ability in Junior Agricultural Staff

	Agricultural Informedness			Explanatory Ability		
	Corn Fertilizers	Grade Cattle Feeding	Pest Control in Coffee/Cotton	Corn Fertilizers	Grade Cattle Feeding	Pest Control in Coffee/Cotton
A. *General Training*						
1. Education: Upper Primary	.14	.22	.13	.15	.17	...
2. Fluency in Swahili	...	...	.22	.20	...	...
B. *Agricultural Training*						
3. Holds Agricultural Certificate	.13	.28	.15	.22	.25	...
4. Was taught with demonstration on topic	...	.26	.16	...	.21	.31
C. *Rebriefing on Topic*						
5. Combined index	.29	.33	.30	.24	.22	.45
D. *Reinforcement*						
6. Specialized in topic	...	...	...	...	.15	.16
7. Informedness of AAO on topic	.15[a]	.40	.16	...	.18	...
8. Number of location staff meetings	...	...	.25	...	...	...
9. Informedness of LAA on topic[b]	.28	...	...	...	.28[a]	...
F. *Total Variance Explained*	27.1%	42.2%	34.9%	24.4%	26.3%	33.3%

NOTE: All figures are partial correlation coefficients, with the other variables for which coefficients are given in the column controlled. All the coefficients given are statistically significant at the .05 level for a one-tailed test of significance.

[a] The Explanatory Ability of the particular supervisor was used in this case.

[b] This item could have been included in the regression equation for three of the other dependent variables at a statistically significant level. It was deliberately excluded because of its contaminating effects. See text.

TABLE 23

The Magnitude of Impact of the Controlled Correlates of Agricultural Informedness and Explanatory Ability in Junior Agricultural Staff

	Agricultural Informedness			Explanatory Ability		
	Corn Fertilizers (0–19)	Grade Cattle Feeding (0–19)	Pest Control in Coffee/Cotton (0–14)	Corn Fertilizers (0–6)	Grade Cattle Feeding (0–6)	Pest Control in Coffee/Cotton (0–6)
A. General Training						
1. Education: Upper Primary (Range 0–1)	1.03 (5.5%)	1.61 (8.1%)	0.65 (5.2%)	0.34 (6.0%)	0.46 (7.0%)	0.15 (2.7%)
2. Fluency in Swahili (Range 0–3)	0.54 (8.7%)	0.37 (5.6%)	0.56 (13.8%)	0.25 (12.9%)	0.17 (7.8%)	0.08 (4.1%)
B. Agricultural Training						
3. Holds Agricultural Certificate (Range 0–1)	0.86 (4.6%)	2.12 (10.6%)	0.71 (5.8%)	0.51 (8.8%)	0.68 (10.3%)	0.16 (2.8%)
4. Was taught with demonstration (Range 0–1)	-0.58 (-3.1%)	2.46 (12.3%)	0.92 (7.4%)	0.14 (2.5%)	0.74 (11.2%)	0.99 (17.5%)
C. Rebriefing on Topic						
5. Combined Index (Range 0–4)	0.88 (18.8%)	0.67 (13.4%)	0.50 (16.2%)	0.17 (11.5%)	0.14 (8.7%)	0.39 (27.4%)
D. Reinforcement						
6. Specialized in topic (Range 0–1)	...	...	...	...	0.46 (7.0%)	0.51 (9.1%)
7. Informedness of AAO on topic (Standardized range 0–8)	0.21[a] (9.1%)	1.04 (41.5%)	0.78 (11.5%)	-0.01[a] (-1.1%)	0.18 (21.3%)	-0.02[a] (-2.6%)

	(1)	(2)	(3)	(4)	(5)	(6)
8. Number of location staff meetings (minus 1 = 0–7)	−0.06 (−2.1%)	0.14 (4.8%)	0.34 (19.3%)	... ...	... ...	... ...
9. Informedness of LAA on topic (Standardized range 0–7)	0.14 (20.5%)	... ...	... ...	0.11[a] (13.2%)	... ...	... ...
E. *Base Level* (Constant)	6.16 (32.9%)	0.74 (3.7%)	2.56 (20.7%)	2.61 (45.3%)	1.75 (26.5%)	2.05 (36.4%)
F. *Total Variance Explained*	29.9%	42.9%	34.9%	24.6%	27.5%	34.0%

NOTE: The range of possible variation for each dependent variable is given at the top of each column. The basic figures given are regression coefficients, which are multiplied times the value of the independent variable and added to the impact of other values in order to give the value of the dependent variable. The percentages given beneath the regression coefficients represent the additional proportion of the total possible amount of the dependent variable which is added by having the particular characteristic. The percentages are calculated by dividing the highest possible score given by the regression equation into the difference in scores given by having the highest and lowest amounts of the particular characteristic.

[a]The Explanatory Ability of the particular supervisor was used in this case.

stand complex Swahili. Although the junior staff use their mother tongues in their homes and in dealings with farmers, official instructions and technical lectures are delivered to them in either English or Swahili. Not all agents are proficient enough in these two languages to understand difficult messages. Most junior staff have problems with advanced English, but even Swahili presents occasional difficulties to a majority of them. Our interviews were conducted in the agents' mother tongue and in it we asked if they ever had difficulty completely understanding technical lectures in Swahili. This reported degree of competence in Swahili is consistently positively correlated with Agricultural Informedness and Explanatory Ability (see table 21, item 2) and can be demonstrated to have independent statistical significance for two of our six dependent variables (table 22). Our best estimate is that an agent who is completely fluent in Swahili will know and be able to explain 8.8 percent more of the total amount of information available on a particular topic than will an agent who often has difficulty with technical lectures in the language. (The estimate of 8.8 percent is the mean of the values for the six dependent variables, as given in table 23. The range of values is 2.8 to 10.6 percent.)[7]

Like fluency in Swahili, education generally increases an individual's linguistic depth and efficiency and hence his decoding ability. Nonetheless the final impact of education on the total amount of information absorbed by junior staff is complicated by its tendency to depress the motivation to receive a message. As can be seen by looking at item 1 in tables 21 and 22, there is a uniform tendency for Agricultural Informedness and Explanatory Ability to rise with education through the upper primary level and then to fall off again with secondary schooling. This curvilinear (arc-shaped) effect is probably the result of the combination of two opposite linear (straight-line) effects. On one hand, the more formal education a member of junior staff has, the weaker his motivation to do his job well, as noted in the previous chapter. On the other hand, the higher the level of his schooling, the greater is his ability to understand and absorb technical messages. Since the amount of information the extension agent actually receives is determined by the interaction between his capacity to decode efficiently and his motivation to use energy on decoding, the information junior staff receive first rises and then falls with education. This combining of two linear effects into a curvilinear one is illustrated with hypothetical data in figure 7. A man with 2.5 years of primary schooling has a capacity of 2 and a motivation of 5, the product of which gives a receiving index of 10. At 5 years of

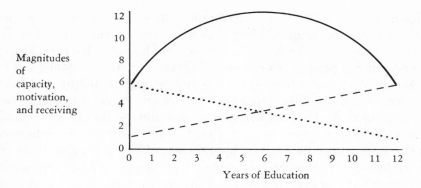

Key:

...... Motivation to receive
-------- Capacity to receive
_____ Actual amount received = motivation x capacity

Fig. 7. The effect of education on junior agricultural staff's reception of technical information.

primary education, capacity is 3, motivation is 4 and the product is 12. Yet a man who spent 10 years in school (3 of which were in secondary school) would have a capacity of 5 and a motivation of 2, which yields a receiving index of 10—the same as that of the agent with only 2.5 years of education.[8] Thus education reaches the peak of its positive effect on Agricultural Informedness and Explanatory Ability with upper primary schooling. Extension agents who have 5 to 7 years of primary education are better informed than both those with only lower primary schooling and those with secondary education. On average they will be 5.8 percent better on Agricultural Informedness and Explanatory Ability than those with more or less education (range 2.7 to 8.1; see table 23).

Specialization on a particular technical topic (item 6 in the tables) also has effects on both decoding capacity and motivation, although this time complementary ones. The agents who specialize in giving advice on grade cattle, coffee, or cotton obviously have a particular motivation to master their specialty area. Furthermore, over the years they have gained a large amount of experience in their area of expertise, which makes it easier for them to understand new information on the subject. The effects of specialization are not apparent for Agricultural Informedness, but they are significant for Explanatory Ability in the agent's area of concentration. The specialist does not seem to be unusually quick in picking up the details of new recommendations but he does enjoy an advantage in understanding how they work and explaining them to the farmer. The greater depth and integration of his knowledge give him an average of 8.0 percent advantage in Explanatory Ability (range 7.0 to 9.1). This does not seem to be as great an advantage as might have been expected for a specialist. The narrowness of an expert's superiority can be explained by the distinct unspecialization of the Kenyan small farmer himself. A farmer who has a grade milk cow is also likely to have coffee trees and to be growing hybrid corn. When an extension agent visits, he will ask about his most pressing problems, whether they are in the agent's area of expertise or not. Thus specialist agents reported to us that fully 24 percent of their farm visits had nothing at all to do with their particular expertise. Similarly, generalist agents gave advice on matters for which there are specialists on 45 percent of their visits. The pressure from farmers causes all agents to try to keep broadly informed, whatever their own speciality. This pressure is also recognized in planning staff training courses, which usually include generalist agents as well as those specializing in the particular crop to be discussed. An organization will have difficulty sustaining very rigid specializations when its clientele does not specialize.

Agricultural Certificate training also improves both motivation and decoding capacity, having a clear, independent, positive effect on five of the six informedness variables (table 22, item 3). The two-year course at an institute of agriculture leading to a Certificate and AA status has a socializing effect upon the new recruits. The institutes' lecturers share a firm conviction of the importance of agricultural extension for the rural development of Kenya. This enthusiasm for their task is communicated in part to their students and strengthens the motivation of the new AA to do a good job. Even more important is the basic training provided by the institutes in the technical details of agriculture. The institutes' courses are one of the sources of an AA's information. But they also give him an understanding of the fundamentals of what is happening when a crop grows. When one knows the basic principles of why something is happening and becomes familiar with the essential processes and vocabulary, it is much easier to learn, relearn, and remember accurately a set of details concerning one of those processes. The details are then not a mass of senseless symbols to be committed to memory but an interrelated and meaningful whole.[9] It is this capacity to understand technical messages that is probably the most important contribution of certificate programs (or any other advanced training). The mass of detailed recommendations needed for advising Kenya's small farmers is simply too great to be remembered for any long period of time. Retraining is a perpetual exercise in an extension organization, as we shall see shortly. The capacity to absorb this retraining quickly and accurately is the basic quality that initial training needs to provide. Possession of an Agricultural Certificate leads to the agent's knowing and being able to explain 7.2 percent more of the total amount of information available on a particular topic (range 2.8 to 10.6). Thus the ministry does not have simply to accept the learning capabilities of its junior staff as a given; its training courses are able to improve their decoders.

The Messages

No matter how strong the motivation of the receiver and how good his decoder, he must be provided with messages if he is to be well informed. Four issues are involved here: the quantity of messages, their timing, the quality of their information, and how well they are transmitted. Virtually all junior staff (99 percent) have been taught the answers to our questions at least once; the great majority have received several additional messages on these subjects as well. The first occasion on which the agent really learned and fully understood a particular set of technical recommendations is especially important, for if that first message is well communicated it will make subsequent communication on that subject much easier. We have reason to believe that those doing this initial teaching are well informed, but there is some variation in how well they are able to transmit their messages. One of the standard methods of improving the quality of information transmission is the use of demonstrations or field visits to emphasize the major points. This technique is a common one in the ministry, particularly in the certificate courses. The method is a valuable one, for those who saw the recommendations concretely demonstrated when they first learned them know 5.5 percent more than do those who did not (range 0.0 to 12.3) and understand 10.4 percent more (range 2.5 to 17.5) (see table 23, item 4). This factor has independent, significant influence on four of the six dependent variables and shows a clear simple correlation with all of them (tables 22 and 21). The effect of demonstrations is weak only with corn fertilizers, where, since it is the staple food crop, the level of common knowledge and experience is quite high. As one would expect, demonstrations are most important when it is difficult for the hearer to visualize what is being taught.

After initially learning a body of information, the agent must receive reminder messages on the subject if his knowledge is to be kept fresh.

The human mind has a large capacity for forgetting and confusing. I once found myself misreporting a major finding of a paper I had written only a year before. Since extension agents are continually talking with farmers about the information they have learned, they remember the messages longer than someone would who rarely used his knowledge. Hence our data demonstrate that the receipt of more than one reminder message in the last year adds nothing at all to the level of an agent's informedness. What a member of junior staff is going to remember, he can recall for a year without further formal messages on the subject. After a year the amount of information the agent can remember does begin to drop off, but still not very dramatically at first—perhaps 2 to 3 percent in the first year and an increasing proportion thereafter. It is therefore important to arrange high quality rebriefing sessions on all current extension topics every two or three years, and it is beneficial to hold them annually. More than one session a year seems undesirable, however, for nothing positive is added by them. Sometimes multiple reminders actually lower the information level. This latter finding is surprising and runs counter to the usual positive effect of three or four repetitions reported in the communications literature.[10] Our particular finding is probably caused by the very low quality of extra rebriefings in the ministry, which actually sometimes provide instruction in wrong information. (This is a problem to which we will return shortly.)

The best rebriefing on a technical topic is one held in the last year, at a residential course in a Farmer Training Centre (FTC), taught by a member of senior staff, and employing a demonstration or field visit. Each of these factors by itself is strongly correlated with all six of our Agricultural Informedness and Explanatory Ability variables (table 21, items 5a–5d). These four characteristics of rebriefing sessions tend to go together, however, so it is impossible to analyze their separate impacts. Instead we have combined them in a single, additive index (item 5e). An extension agent who has received a rebriefing with all four of these desirable characteristics will be 16.0 percentage points higher on the Agricultural Informedness and Explanatory Ability measures than an agent who has had no rebriefing in the last few years (range 8.7 to 27.4; see table 23, item 5). Further experimental research would be necessary to establish the individual significance of these four characteristics. Their combination, however, represents a favorable communication situation. A residential course at an FTC means that the session has been planned in advance. Thus the member of senior staff who teaches it has probably been selected for his particular competence on the subject and has had an opportunity to prepare in advance. His message is therefore reasonably high in quality. If a demonstration or field visit is organized, the agent is exposed to a particularly strong and effective communication channel. The FTC venue also probably reduces the numbers of distractions felt by the junior staff and improves their sense of the importance of the occasion, both of which would add to the quality of their message reception.

We have attended rebriefing sessions under the above favorable conditions and a caveat is necessary about their quality. They are definitely superior to the existing alternatives, but the teachers are still amateurs, not professionals. Some of the communication that takes place in them is decidedly mediocre. In our opinion, a specialized and full-time staff training cadre would do a better job.[11] The Ministry of Agriculture tends to take initial training fairly seriously but to treat rebriefing exercises rather casually. This organizational pattern implies that the decision makers consider basic training far more important to technical competence than rebriefing. Our evidence suggests that this view is mistaken and that the two are approximately equal in importance. Exposure to both certificate training and demonstrations in the initial learning of a set of recommendations gives an agent an average of 15.2 more percentage points on Agricultural Informedness and Explanatory Ability. The comparable figure for rebriefing is 16.0 percent.

The most frequent set of official technical messages received by junior staff are the routine reminders provided by their immediate superiors—the AAO in charge of their division and the AA in charge of their location (LAA). There can be no doubt that these routine reminders have a strong impact on the informedness of extension agents, but the exact nature of their effect is complex. We interviewed all these local supervisors, giving them the same information tests as the junior staff. Thus we know the quality of the messages that they are capable of sending to their subordinates. It is quite clear that the better informed the divisional AAO and the LAA are, the higher are the scores of their staff on Agricultural Informedness and Explanatory Ability (tables 21 and 22, items 7 and 9). One's first temptation is to conclude that a well-informed AAO or LAA raises the level of informedness of his subordinates. But such an interpretation would be mistaken.

Let us first examine the impact of the divisional AAO. Table 24 compares the average Agricultural Informedness performance on our three tests at different hierarchical levels in the ministry. A comparison

TABLE 24
The Degree of Agricultural Informedness at
Different Hierarchical Levels in the Ministry

	Corn Fertilizers (0–19)	Grade Cattle Feeding (0–19)	Pest Control in Cotton/Coffee (0–16)
District/provincial specialists on particular topic[a]	14.3	12.0	6.7
AAOs in charge of divisions	12.3	8.4	3.7
AAs in charge of locations	15.1	11.4	7.6
AAs and JAAs	13.7	8.9	5.2

[a] Those who would be most likely to lecture on the topic at a district level rebriefing session.

of the test scores of the divisional AAOs and of the junior staff (AAs and JAAs) reveals that the AAOs consistently average lower than their subordinates. In fact, in only a quarter of the cases does a divisional AAO score better than the average of his subordinates. Apparently the AAOs both are not sent enough good technical messages and are not well enough motivated to receive them. It is difficult to see how AAOs can be raising the level of junior staff informedness when they actually know less than those they are teaching. It seems clear that although some extension agents may have their Agricultural Informedness and Explanatory Ability increased by technical messages from their AAO, most junior staff can only be confused by such reminders. The damage done by an energetic but poorly informed AAO is particularly great because of the high status given him by his hierarchical position, education, and extensive training. These factors make his messages credible to his subordinates and motivate them to accept what he says.[12] The danger to agent informedness posed by AAOs is somewhat mitigated by their tendency to give their junior staff fewer reminder messages when they themselves are poorly informed $(r = .32)$. This self-restraint is not perfect, however, and our field experience has taught us that senior staff are often completely unaware of how much they have forgotten.

The reinforcement effects of the AAO, both positive and negative, are particularly great on Agricultural Informedness, where the staff of the best informed AAOs score an average 20.7 percentage points higher than those of the least informed AAOs (range 9.1 to 41.5; table 23, item 7). The impact of AAOs on Explanatory Ability is much smaller, an average of 5.9 percent (range 0.0 to 21.3). It probably is easier for a distorted message to lower the level of an agent's Informedness than of his understanding. It is well known that material that is understood is better remembered than material that has no intrinsic meaning. An explanation of why a given set of recommendations is valuable has an internal logic to it, and an agent who understands it will be able to recognize and reject a false one. Many of the details in a set of recommendations have no intrinsic meaning, however, and a misinformed message from a highly credible source can easily confuse the agent. For example, the official recommendations call for the application of two teaspoons of single superphosphate fertilizer to each hybrid corn seed at planting. It would be easy for an agent to become confused if his AAO's instructions were one teaspoon or two tablespoons or some other incorrect detail. It is probably reasonable to conclude that much of the effect of AAOs on Explanatory Ability is due to positive reinforcement, for poor explanations are rejected. The much greater impact of AAOs on Agricultural Informedness, however, is probably caused by the teaching of distorted information (negative reinforcement).

The Agricultural Informedness of the district and provincial specialists who teach special rebriefing courses is well above that of the divisional AAOs and is consistently higher than the junior staff average (table 24). Obviously courses taught by these specialists are much more helpful than reminders given by AAOs. The specialists do not give more than one session in a year to each group of junior staff and any additional reminder sessions that an agent has had will have been provided by his AAO (or possibly his LAA). The inferiority of these supplementary reminders helps to explain our finding that more than one rebriefing a year has a slight negative effect on staff informedness. It is clear that the transmission of information to junior staff along strictly hierarchical lines through the AAOs is decidedly inferior to more direct communication by district and provincial specialists and that, in the present circumstances, the former actually undoes some of the impact of the latter.

The explanation for the correlation between the informedness of the LAA and that of his subordinates is also complex. The LAA is a member of the group that he is supposed to keep informed and receives almost exactly the same set of technical messages that it does. LAAs are never called alone to a rebriefing course at an FTC; they always go with at least some of their AAs and JAAs. Since the LAA and his group of subordinates have basically the same set of information sources, their levels of informedness rise and fall together in response to the frequency and quality of the messages from those sources. Thus the correlation between the LAA scores and those of their staff is largely spurious; the real relationship is between the agent's score and the average one of his group.

A further reason for doubting that LAAs are major formal sources of information for their subordinates is that junior staff almost never report having received a technical reminder from them. The only exception is on hybrid corn, where 16 percent of the agents report messages from LAAs. On this particular topic, the agents with the best informed LAAs are 16.8 percent better in their scores than those with the least informed LAAs (range 13.2 to 20.5). But again, a good deal of this relationship is due to the effect of the group, not of the LAA. Although the LAAs are usually among the very best informed agents in their locations (table 24), those staff who report having had rebriefings only from them generally score lower on our tests (table 21, item 5b). The role of the LAA is probably more of an informal, reinforcing one than of formal instruction. He does not do much initial teaching or rebriefing but he is consulted when his subordinates are unclear on a point. The better informed he is, the more positive will be the effect of his reinforcement.

In our view, the LAA should not be seen as the last link in the ministry's hierarchical information transmission system but instead as a part of the informal information system of the work group. These unofficial communication networks have a noticeably helpful effect on the Agricultural Informedness of their members. As Hovland, Janis, and Kelley point out, cases often arise in which "a person has of necessity to make a judgment on an issue, has no expert opinion or direct reality check as a basis for validating that judgment and, as a consequence, seeks to validate it through agreement with others."[13] Such occasions are frequent for AAs and JAAs, as they are asked to diffuse a wide range of detailed recommendations. Furthermore, at the time of our study, they were not provided with any written references by which to check forgotten details. (This defect has recently been rectified by the issue of a *Crop and Livestock Manual*.) There are two advantages to an agent's being able to validate information with fellow members of junior staff. First, he can ask for advice more easily than he could from an AAO. The status differentials with senior staff are larger than they are among junior staff (chapter 3) and status differences tend to make interaction more difficult.[14] An agent will also be reluctant to expose his ignorance to the AAO who writes his evaluation report. The second and more important advantage to collegial consultation, however, is that the agent has a large number of colleagues available for consultation where he has only one AAO. Most junior staff know more of the details of the standard recommendations than do their AAOs, but virtually all of them have access to at least one other junior staff member who knows more than they do. David Kidd's work in Western Nigeria indicates that extension agents are quite good at evaluating the technical competence of their colleagues.[15] Hence, in a good informal consultation network, junior staff both have advice available fairly readily and know where it can be obtained.

As the amount of junior staff consultation increases, so does the level of their Agricultural Informedness. Location staff meetings increase the opportunities for extension agents to interact with one another. Those locations with eight such meetings a month average at least 7.3 percentage points better on Agricultural Informedness than do those with only one meeting (range 0.0 to 19.3).[16] Increases in the size of the work group is another factor known to decrease the relative amount of possible interaction between individuals.[17] Thus our data show a clear negative correlation between the size of the work group and its average on Grade Cattle Informedness ($r = -.56$, $N = 14$, $p < .025$). Along similar lines, work groups in which job satisfaction is high generally attract more positive feelings from their members and are generally able to have more influence on them.[18] In our case, we find a strong positive correlation between the average satisfaction of the location work group's members and their average Agricultural Informedness ($r = .72$, $N = 14$, $p < .01$).

Conclusions

What general conclusions and policy recommendations can be drawn from the preceding rather detailed analysis? First, information is definitely lost as the number of communication links increases. District and provincial specialists, who receive their information direct from the research stations, are generally noticeably better informed than the divisional AAOs and junior staff, who receive their rebriefing second-hand (table 24). Considerations of organizational efficiency make it inevitable that some of the responsibility for information transmission be delegated to others. But where the delegation of authority usually needs to be encouraged in an organization, tendencies to delegate communications should be kept to the essential minimum.

Second, the addition of another link in a communications chain can be compensated for only if the new transceiver is particularly powerful or is able to fulfill a translation function. In our opening discussion, we acknowledged that an AA would probably get more technical information secondhand in an Institute of Agriculture lecture than he would in a firsthand talk by a Research Officer. This intermediary transceiver must be well motivated if he is to perform his function effectively, however. Junior staff are the final link in the communications chain reaching from the research station to the farmer. Not all extension agents handle this transceiver role equally well. We have seen that those agents with upper primary schooling are better motivated than those with secondary education. As a consequence, they both receive a larger proportion of the messages sent to them (are better informed) and send more messages to the farmers (make more farm visits). The less educated and better motivated extension agent clearly makes the more effective transceiver.

Similarly, provincial and district specialists are better motivated to receive and send messages in their area of expertise than are the generalist divisional AAOs. The wide range of responsibilities of the AAOs precludes their giving quite the degree of attention to technical messages that the specialists do. The consequence is both that the specialists are better informed than the AAOs and that junior staff can learn more from the former than the latter.

A third pair of insights is also related to the issue of motivation. Junior staff feel a good deal of respect for the position and training of the divisional AAOs. As a result they are highly motivated to listen to their technical messages and to take them seriously. Because these AAOs are poorly informed, the result is actually a loss of correct information on the part of the junior staff. One of the important ideas underlined by our study is that bad information is capable of supplanting good if it is contained in the most recent highly credible message received.

The fourth set of points concerns the impact of delay on an organization's information system. It is well known that the more time elapses after one has first received a message, the less of it one will remember. Thus junior and senior staff need to be rebriefed every one to three years. A much less obvious point about the effects of delay is suggested by the summary statements presented in table 25. On Agricultural Informedness, the ability to recall correctly the details of technical recommendations, an information source seems to assume increasing significance the

TABLE 25

The Relative Effects on Junior Staff of the Various Information Channels in the Ministry

	Agricultural Informedness	Explanatory Ability
A. General Education (1,2)	16%	14%
B. Agricultural Training (3,4)	13	18
C. Rebriefing (5)	16	16
D. Divisional AAO (7)	21	6
D'. Work Group (8,9)	14%	4%

NOTE: The percentages represent the average on each of the two sets of measures. The numbers in parentheses refer to the items included, as given in tables 21–23. The differences in the two patterns given here are not statistically significant, as each pattern is based on only three cases.

more recently its message has been received. Thus initial basic training in agriculture accounts for a potential 13 percent; annual rebriefing for 16 percent; and the monthly influence of the divisional AAO for 21 percent. With Explanatory Ability, however, which measures the extent to which the agent can understand and justify a set of recommendations, the relative significance of a source seems to be related to the quality of the instruction it has to offer. The carefully done basic agricultural training adds up to 18 percent; the medium quality rebriefings 16 percent; and the misleading reminders of the divisional AAO for only 6 percent. The Ministry of Agriculture overemphasizes the importance of initial training and underestimates that of rebriefing in its thinking about its communications system. Our findings suggest that this emphasis is particularly damaging for Agricultural Informedness but that it may be somewhat less so for Explanatory Ability.

The foregoing conclusions have definite policy implications and the preliminary results of this research helped to stimulate some improvements in the ministry's communications system. The 1970 Ministry of Agriculture Working Party on Agricultural Extension Services finally recognized that in-service technical training for junior staff is a basic and recurring need.[19] The recommendations of the Working Party set in motion a train of events that has led to the preparation of a manual of crop and livestock recommendations for junior staff and to a more systematic and regular provision for extension agent rebriefing courses.[20] These actions amount to a recognition that the dissemination of technical information to extension staff had been inappropriately left to the district and divisional levels in the past. *The Crop and Livestock Manual* is prepared by the Agricultural Information Centre at the ministry's Nairobi headquarters and the in-service training courses are organized on a provincial basis by the recently appointed Provincial Training Officers.[21]

Nonetheless, we believe that still further improvements need to be made in the information system. It may be worthwhile to discuss a few of these to illustrate how we would suggest approaching this organizational problem. First, the teachers of the rebriefing courses for the JAAs are the AAs assigned to the Farmer Training Centres. Yet no arrangements have been made for systematically providing information to these FTC AA teachers. This reveals a lingering tendency to view the agricultural knowledge that these AAs received on their preliminary certificate courses as permanent additions that require no further replenishing. The forgetting factor is much too important to be ignored in this way. In any case, our research indicates that senior staff members are more effective instructors than AAs. They have greater authority in the eyes of junior staff and they at least have the possibility of direct access to research station information sources. When they specialize on a given topic, AAOs and AOs are likely to be fairly well informed. Furthermore, unlike their colonial predecessors, the present senior staff are African, less distant from the junior staff, and are reported by their subordinates to be better teachers. For these reasons we personally favor the use of a staff training cadre of AAOs or AOs, possibly specialized and operating on a national basis. Others in the Ministry of Agriculture have disagreed with us on this point, however, arguing that FTC AAs are the highest level resource that can be afforded by the ministry at this point.[22] If this latter course of action is accepted, it involves providing systematic and regular training for these FTC AA teachers and accepting the probable information losses associated with one additional link in the communication chain. Whichever strategy is adopted, it is necessary to map out the totality of the information network and to be systematic about arranging communications.

The second flaw that we see in the present system is that no regular provision is being made for updating members of senior staff. Much of the problem here is an assumption that professional officers will read the research literature and keep themselves informed. Such a presupposition may well be valid for the officers who specialize in one technical area over a period of several years, but it is manifestly not supportable for the senior staff in the direct line of hierarchical authority. Most professions provide for annual conferences in which their members can be brought abreast of new developments and refreshed on topics on which they have become rusty. Some such institution to bring together research officers, AOs, and AAOs on a regular, formal basis for serious lectures and an exchange of views would seem essential to extension organizations. The ministry currently holds a few conferences between research officers and field officers, but these are not taken very seriously by those who attend them and are not effective in transmitting information. Formally organized and carefully planned gatherings are probably necessary if the senior staff armor of self-confidence is to be pierced and effective communication is to take place. Only in this way can the dissemination of misinformation by senior staff be arrested.

Finally, the basic problem with the ministry's technical information system is that it shows few signs of organizational planning. The dissemination of recommendations depends on the initiative of the field Agricultural Officer.[23] This task is too weighty for divisional, district, and even provincial officers to handle on their own. They need substantial specialist support from the national level. Equally important, it must be recognized that information is an agricultural input no different from items such as capital and fertilizers. It becomes depleted over time and systematic arrangements have to be made to assure its availability at the right time in the crop cycle. When plans are being made for the introduction or expansion of a new crop or technical innovation, the strategy for moving the necessary information from the research station to the farmer has to be mapped out as carefully as for any other input.

8 Feedback and Innovation

Existing Feedback

The attention we have given to the flow of information from the research station through the extension agent to the farmer may have given the impression that technical communication in the ministry is and should be a one-way process. In fact, feedback (the upward flow of information through the hierarchy and back to the research station) is also important to the success of agricultural extension. No scientist or technical expert, however competent, can anticipate and solve every possible farm problem that might arise in the implementation of his recommendation. Ideally the scientist does his basic research and development work, publishes his recommendations, is told of unanticipated difficulties, and does further research and development to solve the problems and refine the recommendations. Given the wide range of ecological conditions in Kenya and the continuous advancement in agricultural technology, this cycle of research, recommendations, feedback, research, and so on, is never-ending.

Parallel to the feedback and adaptation process involving the research station should be another cycle of feedback and innovation at the local level, run by senior staff of the ministry. The range of local variation in soil types and climate and the length and difficulty of formal research procedures are such that the research stations cannot keep pace with the demands that could be made of them for problem-solving and local adaptations. The ministry expects that the finer adjustments in technical recommendations will be made by Agricultural Officers at the district level. The instruction paper distributed by the Kenya Seed Company with its hybrid corn seed gives the standard national planting and fertilizer recommendations and then advises the farmer to see his District Agricultural Officer for local adaptations.

In practice, however, the processes of feedback and technical innovation have proved weak in the ministry, particularly below the national level. This is not to say that the standard recommendations for hybrid corn, grade cattle, coffee, and so on, are unprofitable when they are not localized. Most of the ministry's recommendations, and certainly all the ones we have been considering in this study, result in profitable increases in yields in almost all the geographical areas for which they are being advocated. Whatever the weaknesses in the ministry's technical innovation process, these recommendations are worth extending to farmers as they now stand. Nonetheless, the profitability of the recommendations could usually be improved, often quite significantly, if local adaptations in them were made.

During the period of our research, we stumbled upon an ideal opportunity for testing the feedback and adaptation capabilities of the extension services at the local level. Through private communication with the national corn research officer, Alistair Allen, we discovered that the standard hybrid corn fertilizer recommendations are suboptimal for the Trans-Nzoia slopes of Mount Elgon. Because of a large amount of volcanic ash in the soil, hybrid corn responds to nitrogen applications but not to phosphorus ones. A soil map of Elgon Location in Bungoma District shows it to have the same soil type as this nearby Trans-Nzoia research area. Thus the standard recommendations of phosphorus and nitrogen need to be replaced in Elgon by ones calling for nitrogen and little, if any, phosphorus. As the Trans-Nzoia research report had had only limited distribution, Elgon Location presented an ideal opportunity to test the adaptive capacities of the field extension services and we specifically included it in our sample for this purpose.

It will be recalled from chapter 2 that we asked all extension staff to answer our technical questions with both the standard recommendations and any changes or innovations that they were using. The extent to which such adaptations were reported then became our measure of Innovativeness. Our field interviews revealed that only two ministry employees in Bungoma District were aware of the desirability of adapting the fertilizer recommendations that we have just outlined. One was the District Agricultural Officer, who had heard of the same research we had. The other was an older Agricultural Assistant who lives in the location next to Elgon, where the same conditions apparently apply, and who had discovered by accident one year that applications of phosphorus are unnecessary. Not only was the capacity for genuine and fruitful innovation limited to this one man but also the feedback structure in the ministry was so defective that he had not convincingly communicated his discovery to his superiors or colleagues. In fact, as we shall see in a moment, it was probably precisely the difficulty of upward communication that was restricting the amount of staff Innovativeness in the area.

Only 13.4 percent of the junior staff in Western Province reported having made any adaptations to any of the standard recommendations about which we asked. Of these, the bulk favored the use of farmyard manure over chemical fertilizers for hybrid corn. Our data show that the propensity to be adaptive does not increase significantly with junior staff education, training, experience, or the like. We find only that junior staff who report that they turn first to a senior staff member when they have a technical question are three times more likely to be Innovative than are those who seek advice primarily from their LAA or junior colleagues. AAs and JAAs generally lack either the technical competence or the self-confidence (or both) to make and act on adaptations in official recommendations themselves. They need advice and reinforcement from an AAO or other senior officer. Thus junior staff cannot be expected to be Innovative unless they profit from relatively easy lines of upward communication.

The relationship between Innovativeness and the free flow of communication is one that has been particularly well documented in the literature on organizations. The presence of hierarchical or status differences in a group tends to inhibit the free exchange of ideas and thus to hinder the process of creative problem solving. Blau and Scott sum up the mass of research on this subject with the statement that only "if leadership does not block but frees the flow of communication . . . [will it] further rather than hinder problem-solving."[1]

Upward communication in an organization like the Ministry of Agriculture is not at all easy to achieve. In chapter 3 we discussed the social barriers that exist between junior and senior staff. But the geographical dispersion of the extension services presents an even more formidable obstacle. Even the LAAs, who suffer no social barriers with their subordinates, see each of their AAs and JAAs only 3.1 times a month.[2] The AAOs' contact with their junior staff is even less—an average of 2.4—and these contacts include staff meetings and paydays, where the contact is usually formalized and superficial.[3] Divisional AAOs do report seeing their District Agricultural Officer (DAO) 5.3 times a month, but then the DAO has many fewer AAOs to supervise than does the AAO of AAs and JAAs. Divisional AAOs have only 1.3 contacts per month with any of the provincial level specialists. The nature of the work makes it difficult to increase the quantity of personal contacts between hierarchical levels and particularly between senior staff and subordinates. Only the quality of the exchanges is really subject to improvement. Thus the AAOs who remain genuinely open to questioning and consultation by their staff when they are able to be in contact are the ones who stimulate feedback and Innovativeness. When we discuss a general structure for the supervision of junior staff in chapter 10, we will indicate ways in which such problem-solving interactions between agents and their supervisors can be made a routine part of the extension system.

Nonetheless, it would be wrong to consider the junior staff as the center of the local adaptation process. They are an invaluable source of information about local crop experiences and problems but their agricultural training is not deep enough for them to be trusted with altering technical recommendations. The ministry quite properly considers Innovativeness to be a senior staff function. We found that a good proportion of the senior staff in Western Province are aware of the problems in the recommendations and are toying with the possibility of small adaptations in them. Nonetheless, they seldom seem to promulgate any to their subordinates, and they do not seem to be collecting the kind of crop performance information that would show them precisely what sorts of adaptations are desirable. The safe course is felt to be the bureaucratic one: to apply the hierarchically approved programs. At the moment most officers feel that the localization of recommendations must come from the research stations. They limit their role to the occasional intimation of problems to research personnel when they are asked to do so.

Like most of us with an academic orientation, the research officers are very reluctant to make any recommendations other than those they have scientifically established. For example, we have examined some of the files at the National Agricultural Research Station in Kitale, and we are persuaded that the information exists from which rough estimates could be made concerning the appropriate fertilizer requirements for corn in different areas of the country. Rather than conjecture on the variations, the station issues a single set of recommendations based on what they know to be appropriate on their research farm. Scientists are unlikely to make educated guesses unless field officers put considerable pressure on them to do so or start making them themselves. This certainly needs to be done, for a good guess about a way to increase the profitability of a farmer's operation this season is more valuable to him than a carefully proven recommendation five years later.

We did encounter one provincial officer who had initiated a localization of the recommendations applicable to his specialty. He was doing so with the national head of his technical division and was in frequent contact with him about the adaptations that were needed. It is precisely such hierarchical openness to feedback and encouragement of Innovativeness that is required at all levels in order for the processes of adaptation to take place.

Diagnosis

The weakness of feedback and innovation at the field level in the Kenya Ministry of Agriculture derives from a failure to appreciate their potential contribution. The diagnosis of problems in farm systems is viewed as a highly technical process and is therefore reserved to centrally provided consultants. Whether soil scientists and entomologists from a research station or economists and sociologists from the U.N., the consultants are expected to collect their own data, analyze them with sophisticated methodologies, and produce recommendations for new directions in local development. Such outside reviews of an agricultural system are necessary when serious problems are encountered or when large capital investments are contemplated. But these exercises are expensive and necessarily either infrequent or superficial. Feedback and innovation are needed not only on the grand contours of farm technology and economics but also on the intricate details. These fine points cannot sustain the attention of a consultant and must be dealt with by local farmers and staff if they are to receive the regular attention they deserve. There is a place for diagnosis and adaptation at the local level.

The first step in promoting a local staff role in the adaptation of extension recommendations and programs is to recognize that feedback must begin outside the bureaucracy. It is the farmer whom the shoe pinches and whose experience must be tapped. The farmer must be consulted; but it is difficult to do this meaningfully or well. Meetings with groups of farmers are not likely to be very useful unless they are carefully planned.[4] Small farmers see such gatherings as forums in which to appeal for government funds or assistance, and problems that are not perceived as falling within that framework are unlikely to be mentioned. Thus farmers will petition about the need for roads, the high price of fertilizer, or the shortage of credit, but not about the extension recommendations that require labor at times when most of them have none to spare. The possibility of outside assistance in solving *technical* problems will tend to be overlooked in these forums and the demands for government expenditure are most likely to be unrealistic and unjustified by any reference to corresponding increases in local economic productivity.

A second difficulty with meetings is that the problems presented tend to be those of the community's leaders, while those of its poorer majority are often ignored. This is not to say that local representative councils and farmers' associations are dispensable in agricultural planning. Their contributions can be very great indeed in setting priorities, providing political support, mobilizing resources, and pointing to unanticipated problems. But the agricultural officer himself has to be able to bring to these meetings clear ideas of what national resources are realistically available and of the new production and technological opportunities that exist. These matters demand knowledge beyond the small farmer's range of experience and the officer has to be able to argue for the unseen possibilities and against the inflated hopes in order to get a useful discussion. To do so he will need an intimate knowledge of the local agricultural systems. This returns us to the problem of how the farmers' experience is to be tapped.

A brief interview with an individual farmer also has its limitations. As Philip Mbithi has argued forcefully and well, rural peoples have tremendous capacities for adaptation and development within their own frame of reference. The outside professional with a different frame of reference may not understand the true meaning of what he is being told. The small farmer is likely to describe his problems in different terms from those that would enable the expert to recognize them and prescribe the cure.[5] He also will not fully understand or have articulated in his own mind the nature of his farming problems. Part of the reason he needs agricultural experts is to tell him what the difficulty is, when he only intuits that something does not work.

To be meaningful the expert's interaction with the small farmer has to be intense and extended. This is illustrated by what Leonard Joy calls

> thinking it through with the farmers. I remember a particularly instructive two days with senior extension officers in Bengal where first of all a farmer explained what he did, day by day through the cropping calendar, and then the extension officer told him what he should be doing and we then tried to work this out day by day through the cropping calendar. In this process it became absolutely clear that the farmer could not sensibly do what he was being advised.

This was, of course, because of timing constraints. In other situations, however, considerations of uncertainty would be brought out or perhaps problems of access to resources and so on.[6]

For the local officer the process of consultation can and should be even more extended, and thereby more meaningful. One of the best ways of doing this is for the field officer to select annually a small sample of farmers with whom he will consult often and whose farming he will observe closely throughout the cropping cycle. The very process of close involvement with the whole farm experience of a set of small farmers makes the trained officer forcefully aware of a series of technological, economic, and sociological problems. As the senior staff become conscious of these problems, they can begin to seek their solution.

We suggest the intense diagnosis of a small sample of farms. The benefit of the exercise flows not from intensity alone but also from the representativeness of the sample. We found in Western Kenya that most officers do have a few farms with which they keep in touch and which they consider (perhaps unconsciously) to be typical. The problem is that these farms are actually atypical, often dramatically so. An extreme example is the officer in a district of 750,000 who told us that there was a lot of interest in pig husbandry in the area and that the farmers engaged in it had achieved high standards. We subsequently discovered that only two private farms in the whole district had pigs and that the one described to us belonged to a resident senior civil servant. More typical was the widespread comment by senior staff that their most junior extension agents were now less well informed than the majority of the farmers. Yet our interviews with agents and farmers indicate that probably only 4 percent of the small farmers were better informed on corn technology than the least knowledgeable of the junior staff.[7] In the absence of careful controls, extension staff develop quite inflated impressions of the quality of farming in their areas and then gear their extension programs to problems that are irrelevant to the average farmer.

In order to assure the representativeness of samples of small farmers, we have become strong advocates of random selection. The bias of government staff toward wealthier, more progressive farmers is so ingrained (as will be seen in chapter 9) that only a systematically drawn probability sample is likely to bring field officers face-to-face with reality. There are qualifications that have to be made in our advocacy of systematic samples. Sampling techniques for large rural areas are quite complicated and are unlikely to be readily mastered by field officers.

Furthermore, the resulting sample for, say, a division would commit the officer to such a widely dispersed group of farmers that he would waste a great deal of time and fuel on travel. However, an officer can relatively easily select a small area (a square kilometer) on a rotating annual basis for his particular attention. The area might be chosen for a suspected agricultural problem or opportunity. Once the area is identified, the officer can have a junior staff member prepare a list of all the farmers in it and draw numbers from a hat to select a small sample of ten to twenty, which he can work with intensively that year. If he is interested in farms with a particular type of crop or production possibility, the list from which the sample is drawn could consist of the farms with that characteristic only. The drawing of numbers should probably be done at a public meeting in the area so that everyone understands how the selection is made.[8]

Ten to twenty intensive farm case studies will not lead to a set of descriptive statistics on agriculture in the division or to a quantitative proof of a causal relationship. These are not the types of data that are needed. What is required is that the extension officer begin to develop insight into the farmers' problems in his area and that these perceptions apply to typical conditions in that agricultural system. The case-study diagnosis will generate understanding; the small random sample will assure that a range of farm types is confronted, with an emphasis on the most common. The insights gained in this way will be supplemented of course by the problems that the officer's junior staff are bringing to him about the farms on which they are working.

Many of the problems and opportunities on the farms confronted will be readily identified by the field extension officer; some will be beyond his competence. A feedback process that reaches beyond the individual member of senior staff and includes his peers and national specialists is necessary if the opportunities for problem solving and sensitivity provided by the case studies are to be fully exploited. The literature on organizations is particularly clear on the importance of a free flow of consultation among *equals* in problem-solving behavior.[9] This would suggest the desirability of local groups of ten to twenty field officers getting together regularly to discuss their problems and successes. This process might be encouraged by having monthly seminars in which each officer in turn presents detailed information on his difficult or interesting case studies and experiments. The object would not be to offer firm conclusions but for the officer to present the data needed for a creative, problem-solving exchange. The formalization of consultation in this way would go a long way toward promoting a sense of professional collegiality among field staff.

The problems that local officers find especially difficult or interesting should be passed on to national specialists for their advice, further consideration, and research. The officers at Kenya's national corn research station tour the provinces annually, holding seminars with senior staff in which new research results are presented and problems with corn production are sought out. This kind of itinerant consultation is essential in all technical areas. For the presentation of problems to be truly fruitful, however, it has to be prepared by local officers who have been engaged in intensive diagnosis and collegial discussion.

Feedback and adaptation then should be seen as occurring on multiple, interacting levels. Not only does the competence to solve various types of problems exist at differing levels; the solutions may be different depending on the frame of reference. Paul Devitt points out that

> An ideal structure ... would also interpret problems differently according to the perspective from each level. For example, a problem of low grain production may be seen as due to low prices at the village level, poor road communications at the provincial level, and of import tariffs at the national level. The diagnosis of the problem is therefore different at each level and so is the prescription for a remedy. But such a structure can only function effectively if each level is doing the work appropriate to it. Institutions at different levels should communicate and cooperate, but they cannot substitute for one another.[10]

Only if close involvement with poorer farmers, professional collegiality, and free local-national interaction are promoted is creative problem solving going to exist.

Experimentation

Diagnosis is only the first step to innovation. No matter how sophisticated the analysis has been, a solution must be tried experimentally before it can merit confidence. The causal networks in biological, economic, and sociological systems are so complex that we can never be certain of the true constraint on a particular development process until we have seen what happens when we remove it. Field trials are essential to successful extension work.[11] But the current popularity of itinerant, short-term consultants has led to the eclipse of these experiments.

Senior field staff must play a central role if experimental work is to assume its proper place.

Before we can spell out the innovative role of field officers, we must acknowledge the existence of levels of magnitude and complexity in experimental work. Many variables demand national decisions before they can be changed and are obviously beyond the scope of local staff. Other problems are so complex that intricate experimental designs and high-level expertise are required for their solution—plant breeding being the prime example. Field officers can experiment only within the parameters set by the larger technological, economic, and political systems in which they work. Nonetheless there is almost always a considerable amount of room for minor but significant adaptations within these limits. Small-scale local experiments also offer the opportunity to demonstrate the need to change the parameters of the existing systems. In the first category of minor adaptations are matters such as the exact fertilizer mix appropriate to local soil types; the relative returns to labor of different crop possibilities; the feasibility of new crops, varieties, or animals under local conditions; local readiness for cooperative marketing groups; and the desirability of using different social groups in extension for a given community. Answers to these sorts of questions can be provided only locally, for they involve details that cannot be provided economically by high-level experts. Experiments such as these follow naturally from the diagnosis of sample farms and it probably would be ideal if they were carried out on the same farms in the season following the diagnosis. The resulting changes would be better understood by the officer because of his previous experience with the farms and would provide some returns to the cultivators who had put up with his time-consuming questions the previous season.

The second category of experimentation, preparatory to national action, involves a different type of trial. The feasibility of a major irrigation scheme may be demonstrated first by an officer carrying out inexpensive, small-scale trials in the area. (This is the way the Mwea Scheme in Eastern Kenya began.)[12] The economic and social benefits of land reform may be demonstrated by senior local staff who help landless farmers take over an abandoned estate temporarily and unofficially. A need for the alteration of national prices might be demonstrated by providing an indirect subsidy to a small group of producers and seeing how great their response is. One imaginative Kenyan Assistant Agricultural Officer in Eastern Province was asked to promote cotton in an area where the profit margins were too low. He tied his cotton campaign to the introduction of an extremely profitable new variety of beans. By marketing the two together and pooling the returns, he made the cotton appear more profitable than it was, leading to a short-term spurt in the growing of cotton.[13] Obviously such an arrangement neither could nor should be sustained, but it demonstrated that the constraint on cotton adoption in the area was prices and not some sociological or technical problem. When made known through upward feedback, such a demonstration focuses the attention of national policymakers on the problems they must solve.

Once diagnostic and experimental work has been defined as a central and essential part of the responsibilities of senior field staff, a set of itinerant national specialists will need to advise and consult them. Information gleaned from the diagnostic and experimental work of local officers will form one of the bases for setting national research priorities and for identifying the developmental constraints that require national policy attention. The field officers will need advice as well. Their diagnostic and experimental designs may need to be improved and their analyses of their experiences refined. Peer group discussions will help, but specialists who offer supporting advice will enable all concerned to learn more.

The combination of local diagnosis followed by adaptive and preparatory experiments is essential to good extension. But do the personnel exist who can perform these functions? Our Kenyan experience convinces us that those who hold diplomas in agriculture can do the elementary diagnosis and experimentation needed, and we therefore believe these functions should be a part of the routine duties of all senior staff. When officers who are used to following orders are first asked to suggest new directions for rural development, their responses tend to be unimaginative and lacking in insight, as was shown in the early stages of the Kenyan Special Rural Development Programme. But this is because they have not been given the time and encouragement to build up diagnostic insights and experimental experience. Conversations with these same supposedly unimaginative officers demonstrate that they have many good but half-formed ideas and only need support to go on and develop them. Obviously officers with only intermediate technical training will make mistakes; but so do those who have Ph.D.s. The amounts of locally specific knowledge that can be gained through experimentation are worth the risks of error. Also, officers who have experimented with something themselves and seen it fail will not duplicate their mistakes. Classroom precepts are quickly forgotten and nonexperimental action in the field, where the link between action and results is unclear, permits the field officers to repeat their errors over and over again. Good quality support from specialists should make it possible to avoid the grossest errors. As Guy Hunter has put it, "The barefoot economist or agronomist or engineer is more probably two men—a retrained junior field man with access, for advice and supervision, to an experienced professional."[14]

Another potential difficulty with experimentation by local staff is that it requires time to be brought to fruition, and in Kenya the transfer rates for senior staff have been relatively rapid. Sound extension work requires that men spend at least three years in a post and personnel policies have to be directed to that end. But turnover can only be slowed; it cannot be eliminated. When turnover does occur, there is the danger of discontinuity in the whole direction of local agricultural development. This is another reason, then, for stressing the element of collegiality among the senior staff working in an area. If diagnostic findings are discussed by the peer group and experiments grow out of collective decisions, the senior staff as a social unit can provide continuity where the individual officer does not. Collective learning experiences have a much more enduring impact on a locality than do the personal ones of particular officers.

Finally, our stress on the importance of local experimentation implies a certain element of deconcentration in the administrative system. Officers will not work on new directions for local development if their experience will never be incorporated in the programs they administer. Furthermore, although a minor experiment requires only modest resources (probably no more than $2,000), an officer must have discretionary authority over some small funds if he is to proceed. Colonial agricultural officers usually could command such modest funds, but officers in independent Kenya have found their votes for recurrent expenditures ever more seriously constrained. (In contrast, finances for major development initiatives are now more readily available.) It is important the the treasury recognizes that local experiments can produce better information at a lower cost than can expatriate consultants and that it therefore consents to the creation of small discretionary research votes. In addition, central planners must consult and pay attention to the experience of local staff in devising their programs. Of course a more general deconcentration of financial decision making, such as Tanzania's locally controlled, medium-sized Regional Development Fund, would be an even better solution. But an experimental role can be sustained with only a moderate degree of discretionary authority and a serious effort to stimulate upward communication.

Summary

National agricultural programs and technical recommendations must be adapted to local conditions if they are to be fully effective and appropriate. This task cannot possibly be performed by the agricultural research establishment or the nationally based specialists, who are expensive and already overcommitted. The present orientation of Kenyan field staff toward centrally promulgated recommendations means that the amount of local adaptation and innovation is quite limited at the moment. Ultimately, senior field staff must discharge this function if it is to be done at all. We noted that hierarchical formalism greatly hampers upward communication of problems and insights and that deemphasis of status differentials tends to encourage such feedback and accompanying innovations. The ways that feedback from junior staff can be promoted will be discussed in chapter 10 in the context of a general supervisory system. The adaptive role of senior field staff should be promoted by making diagnostic and experimental work a part of their routine duties. We suggested that each year an officer do intensive case studies on a small, randomly selected sample of farms in order to improve his sensitivity to the opportunities and constraints in the local agricultural system. Experimental field trials could be simultaneously conducted on the previous year's sample of farms to test out insights on new directions and modifications in the local extension program. A strong system of local senior staff discussion groups and itinerant specialist advisers was recommended to support such diagnostic and experimental work. We are convinced that the result would be much greater relevance, dynamism, and impact for the extension services.

9 The Distribution of Benefits

The Extent of Inequality in Visit Effort

The 1970s have brought a very marked change in the outlook of international development agencies. The emphasis is now not only on agricultural growth, but on the alleviation of rural poverty as well. In the 1950s the accentuation of rural inequality was not just accepted; it was encouraged. Kenya's colonial government was quite explicit about it: "Former government policy will be reversed and able, energetic or rich Africans will be able to acquire more land and bad or poor farmers less, creating a landed and a landless class. This is a normal step in the evolution of a country."[1] During the 1960s it was naively assumed that the rural poor would benefit from general rural development initiatives. The lessons of India's Green Revolution and the food crises of the early 1970s have made it clear that the forces producing rural inequality require direct and careful attention.[2] In this chapter, we demonstrate that current extension practice in Kenya is accentuating the gap between the wealthier minority and the poor majority of small farmers. We also identify the policies and pressures that are producing this sometimes unintended effect so that we can find measures to counteract it.

The first step in assessing the distribution of agricultural extension benefits is to establish the nature of the farm population being serviced.[3] To do this, we have used a portion of a large survey of small farmers conducted by the Agricultural Statistics Section of the Ministry of Finance and Planning during the 1970 long rains.[4] The survey gives us detailed information on 637 randomly selected farmers in Western Province.[5] From an analysis of the Agricultural Statistics data, we can gain an accurate picture of the distribution of various easily identified farm characteristics. The growing of hybrid corn is one of these. Corn is the basic food for the great majority of people in Western Province, and hybrid corn is a relatively recent but well-established agricultural innovation in the area. The package of hybrid seed and fertilizers was introduced in the province in 1963, and hybrid corn (with or without chemical fertilizers) is now grown by 48 percent of the farmers there. The return on the use of the hybrid and fertilizer package varies, but it is not likely to be less than a 100 percent net profit over a farmer's extra cash investment. Thus, a farmer is likely to have accepted hybrid corn if he has been innovative over the last few years. Nonetheless, the use of hybrid seed varies from an estimated high of 80 percent of the farms in Kimilili (Bungoma) and Lurambi (Kakamega) Divisions, where land holdings are large and corn is a major market crop, to a low of 4 percent in the Central and Southern Divisions of Busia, where soil and climate are less favorable and where cassava competes with corn as a food staple.

Different cash-producing farm enterprises are appropriate to each of the ecological zones in the province, and the profitability of these enterprises varies considerably. Grade dairy cows have a very high return on investment, whereas the profitability of cotton is relatively low. The prices on the robusta type of coffee (but not the arabica) are so low now that many owners of these trees do not consider it profitable to care for them or to harvest the berries. Nonetheless, ownership of one of these farm enterprises does indicate that the farmer has had investment funds available at some point in the past and that he is now or was once deriving a cash income from his produce. This marks him as being above average in wealth in what is still predominately a subsistence economy. Farmers with such cash-producing farm enterprises constitute 15 percent of the total in Western Province.

We can define a progressive farmer as one who both uses hybrid corn and has one or more cash-producing farm enterprises. Only 10 percent of the farmers in Western Province meet these two criteria. This definition approximates the minimum behavior that agricultural staff in Western Province expect of what they call a progressive farmer. Such a farmer probably has been innovative over a fair period of time, has access to small amounts of capital, and is well-to-do relative to his neighbors. Conversely, we will define a man who has neither hybrid corn nor a cash enterprise as a noninnovator. In Western Province, 47 percent of the farmers fall into this category. For these farmers, the adoption of new farming methods is not a habit and access to investment capital is often a problem (see table 26).

Of course, it does not follow automatically from a farmer's being progressive that he is relatively rich. For this reason the Agricultural Statistics Section's survey data on cattle holdings is particularly interesting. In the past cattle were overwhelmingly the symbol and substance of wealth in rural Kenya. Although this traditional attachment to cattle has diminished in Western Province, a Luhya's wealth is still likely to be

TABLE 26
The Distribution of Agricultural Enterprises among Farms in Western Province

	Have Cash Farm Enterprise	No Cash Farm Enterprise	Totals
Have hybrid corn	10%	38%	48%
Have no hybrid corn	5%	47%	52%
Totals	15%	85%	100%

NOTE: Based on a weighted sample of 637 farms. Excludes Northern Division, Busia, and the settlement schemes in Bungoma and Kakamega Districts. Data collected by the Agricultural Statistics Section of the Ministry of Finance and Planning during the 1970 long rains.

A cash farm enterprise is defined as one of the following: grade cattle, coffee, cotton, or tea.

reflected in his livestock holdings. Thus, it is interesting to note that those who grow hybrid corn in Western Province are twice as likely to have five or more cattle as those who do not grow it. (For the purposes of this exercise, one grade cow is counted as equal to two local cattle, the difference in their market value.) Furthermore, those whom we have defined as progressive farmers are one-eighth as likely to have no cattle as those whom we have labeled noninnovative (see table 27). Thus we see a fairly clear relationship between progressiveness and wealth. The only exception is that small category of farmers who have adopted a cash crop but not hybrid corn. These are very much like the poor farmers in their livestock wealth. The bulk of farmers in this category raise cotton in Busia. As cotton seed is provided free to the grower, it is the one cash crop that does not require a capital investment to plant and hence is accessible to the poor. Unfortunately, cash investment is required for insecticides if the plant is to produce good yields, so most farmers in this category will be disappointed by their harvest and remain poor.[6]

Having identified the proportions of farmers who can be called progressive and noninnovative, we now have a baseline from which to compare the actual distribution of agricultural extension services. We have already pointed out that the basic technique of extension in Western Province is visits to individual farmers. In our interviews we asked each staff member who works in direct contact with farmers to name for us all the farmers to whom he had paid extension visits in the previous week. For each of these farmers we then inquired whether he grew hybrid corn and whether he had a cash farm enterprise. In the province as a whole, the average extension agent spends 57 percent of

TABLE 27
The Cattle Holdings of Farmers in Western Province

Number of Cattle[a]	Noninnovative Farmers		Middle Farmers	Progressive Farmers	All Farmers	
	Neither Cash Crop nor Hybrid Corn	Cash Crop but No Hybrid Corn	Hybrid Corn but No Cash Crop	Hybrid Corn and a Cash Crop		
0	198 (63%)	32 (67%)	77 (35%)	4 (8%)	311	(49%)
1–4	56 (18%)	7 (14%)	54 (24%)	19 (37%)	136	(21%)
5 or more	58 (19%)	9 (19%)	91 (41%)	29 (56%)	187	(30%)
Total	312 (100%)	48 (100%)	222 (100%)	52 (100%)[b]	634	(100%)

aThe very few grade cattle found in the sample were counted as the equivalent of two local cattle, the approximate difference in their sale values.

bThe apparent total of 101% derives from rounding errors.

his visits with progressive farmers (who are 10 percent of all farmers) and 6 percent of his visits with noninnovative ones (47 percent of the total). If extension were distributed perfectly equally, one visit would be made to every other farm each year.[7] Yet our data show that the average progressive receives approximately 2.91 visits a year, while the middle farmer gets 0.44, and the noninnovator 0.07 (see figure 8). Thus extension attention is very greatly skewed in favor of the more progressive and wealthier farmers. This concentration on progressive farmers is achieved largely at the expense of the noninnovative ones. Farmers who have either hybrid corn or a cash crop but not both are 43 percent of the total and extension agents devote an average of 37 percent of their visits to them. A farmer in this middle category, who has shown some innovative drive, has about one-seventh the chance of a progressive farmer of receiving an extension visit. But his odds are still 6.5 times those of a noninnovative farmer, who has one forty-fourth the chance of a progressive farmer (see figure 8). This Progressiveness Skew is not peculiar to Western Province or to Kenya. Studies by Joseph Ascroft and by Solomon Njooro have established its existence in central Kenya, and H. U. E. Thoden van Velzen has documented it in Rungwe, Tanzania.[8] In fact, this problem is probably general to all agricultural extension services in the world and only the degree of it varies.[9]

Policy Determinants of Inequality

What are the causes of the emphasis placed on progressive farmers in extension work? The most important factor is the strategy that agents have consciously and openly adopted for their work. In a Tanzania-wide opinion survey of farmer-contract extension agents, R. G. Saylor found that 87 percent agree with the statement, "If I worked most of the time with a few of the better farmers, I would get better results." This opinion was expressed although it runs contrary to the official policy of the Tanzanian Ministry of Agriculture, Food and Co-Operatives.[10] That junior staff should support so openly a strategy disapproved of by their minister indicates that they believe there are strong, legitimate arguments behind it. The progressive farmer strategy enjoys deep support among extension professionals at all levels in East Africa and is an important determinant of their behavior.

Nonetheless, it may be that the strategy is mistaken. To investigate this possibility, we need to examine the two major justifications for the progressive farmer approach. The first arises from the diffusion of

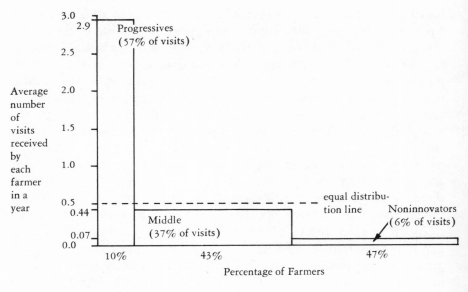

Fig. 8. The distribution of agricultural extension visits to farmers

innovations school. Progressive farmers are not only those most receptive to agricultural change, they also represent the informal leadership of their communities on technological matters. Innovations proved on their farms will diffuse to the other farmers in the area through a natural process of social communication. Therefore, say its advocates, extension agent concentration on progressive farmers simply represents a highly efficient technique for achieving a wide distribution of benefits to the entire farming community.[11]

The second set of arguments is most frequently offered by economists. Here, extension services are seen as only one of a number of agricultural inputs and the concern is for their most economic use. Progressive farmers have a number of characteristics that make them the most efficient target of agricultural extension: they are psychologically predisposed to change and so require less persuasion; they have access to the other inputs necessary for innovation (especially capital) and thus are quicker to change once they are convinced; in East Africa they typically own larger farms and the adopted innovation will therefore be applied to larger acreage. For all these reasons, more agricultural output will be achieved for the average visit to a progressive than will be gained per visit to other farmers. Thus the economic arguments for the progressive farmer strategy do not depend upon its achieving a wide distribution of benefits. Extension is conceived of as a tool for economic growth, not social welfare.[12]

The wide distribution of benefits promised by the diffusion of innovations argument depends on two conditions that often are not met in Western Kenya: (1) the proposed innovation will remain available for eventual adoption by all or most farmers; and (2) there are not significant social barriers to the communication of agricultural practices from progressives to others. Hybrid corn has been the model for most diffusion of innovations thinking.[13] As corn is the staple food crop in East Africa, it is likely that hybrid varieties will ultimately be grown on most small farms in the region. But this wide potential spread is not a common characteristic of agricultural innovations. Coffee and tea are more typical of the new crops offered to farmers in Africa. Only a small proportion of all farmers will ever grow either. When African growers in Kenya were finally permitted to raise coffee, the innovation began to spread rapidly, the market became saturated, and new plantings were prohibited. Hence progressives were not just the first to adopt this innovation; they were the only ones. Tea illustrates a slightly different pattern. When the crop was first introduced to small holders in Kenya, the price of cuttings was subsidized and the minimum area to be planted was a quarter acre. Small-holder tea is still being expanded, but since the innovation is established, cuttings are now being sold at their full cost and the minimum planting is one acre.[14] Thus the investment was made easy and was subsidized for the relatively well-to-do progressives; it is not for the poorer mass of farmers. The conclusion is that access to extension services and the early adoption of an innovation is not simply a temporary advantage; it often represents a permanent gain in the basic profitability of the progressive's farm relative to that of his neighbor. Similar conclusions on permanent income disadvantages to late, less well off adopters have been reached in investigations on the diffusion of miracle rice in India.[15]

It is true that government actions denied the benefits of coffee and tea production to the poorer farmers. Nonetheless, these policy decisions were related to the economics of agricultural production, and the free market would have produced a similar effect eventually. As the supply of a crop expands, its price will drop and so will the profitability of growing it. The early adopter reaps the largest returns; the late one the lowest. If the crop requires any long-term capital investments (in trees, machinery, buildings, and the like), the early adopter will be able to pay those initial costs out of the higher returns he makes in the early low-supply market. When the supply expands and the prices drop, he will have to meet only his recurrent and not his initial investment costs out of his sales revenues. Thus his marginal costs will be lower than those of a farmer just beginning on the crop, and his profits will be correspondingly greater. The farmer who adopts the crop and makes his initial investments relatively late may well take a net loss on it at the same time that early adopters are still making a profit. A government decision to prohibit expanded planting of a crop with a saturated world market, such as coffee, actually may save the potential late adopters from net losses (as well as protecting the profits of the early ones), for it is anticipating the operation of the free market. Nonetheless, even if such a decision is benign toward potential late adopters, it does not alter the benefits gained by the early ones on this type of crop. It seems to us that there are only three situations in which the economics of production will not give a permanent advantage to the early adopter: (1) when the crop is grown largely for home consumption; (2) when no long-term capital investments are necessary for efficient production; or (3) when the crop is sufficiently specialized to one region that it will be profitable for virtually all farmers in that area to grow it, however late they adopt it. Only a few crops are in one of these three categories. Thus the diffusion of innovations school is wrong in its assumption of widespread adoptability for most agricultural extension projects.

The second major condition upon which the diffusion of innovations depends is a free flow of agricultural information in the farming community. We have good reasons to believe that this condition is often unfulfilled in Western Kenya, although our data base here is too weak for our discussion to be conclusive. Early in our research, we interviewed a small random sample of farmers in the Vihiga Division of Western Province. We did find that at least some farmers who have no personal contact with extension workers are getting new agricultural information from other farmers who do have such contacts. Thus diffusion of new information does occur. Nonetheless, it is not enough to know that some information about an agricultural innovation is being passed from farmer to farmer. We must know how much. One study of information diffusion in India showed that the farmers who first learned about an innovation passed on only 28 percent of what they knew to other farmers. As the first learners themselves missed much of what had been taught, the farmers who got their information secondhand from their neighbors learned only 14 percent of the points the Ministry of

Agriculture had been attempting to communicate to them.[16] This is a rather slim information base for farm decision making. Furthermore, our interviews in Kenya's Western Province suggest that what is forgotten in the diffusion of information from one farmer to another is vital for his deciding whether or not to adopt. The several farmers with whom we spoke who had heard of a hybrid corn innovation only through other farmers had no idea whether the change would increase yields. Without yield information, an innovation discussed with others carries little conviction and is unlikely to be adopted.[17] Yet it is uncommon for farmers in Western Province to reveal freely the amount of profit they have made from an innovation. Publicized income differentials may give rise to a higher tax assessment, increased social obligations, jealousies, or even, rarely, accusations of witchcraft. Thus most Luhya farmers probably require a new institutional context in which discussions on profitability are expected before they will talk readily about yields.[18] Visits from professional agricultural change agents and meetings organized by them are the main social settings in which the crucial question of returns will be treated in many parts of Western Province. The informal channels for the dissemination of agricultural information are therefore not as strong as the diffusion of innovations theory presupposes. Of course a good, new agricultural practice will still spread despite the weakness of the informal information system, but general acceptance will be slower than would otherwise be the case.

If there are barriers to the free, informal communication of agricultural information, it would seem undesirable to apply a strategy of working almost exclusively with progressives. When a broadly applicable innovation, such as hybrid corn, is first being introduced, it could be thought wise to begin with the progressive farmers as those most able and willing to take the associated risks. Once the new practice has gained a foothold, however, it would be rational to shift attention toward the less innovative farmers to speed the spread of adoption. It can be inferred from table 28 that such a strategic change in focus does not occur. General extension agents, who carry the burden of work on hybrid corn, give only a tiny proportion of their time to the half of the province's farmers who do not grow it. Further, they devote at least as many visits to progressives as do their specialist colleagues, who would have much more justification for working with an advanced clientele. Presumably the general agents are trying to achieve improvements in the technical standard of cultivation on the farms of the adopters rather

TABLE 28
The Average Percentage of Extension Visits to
Farmers by Agents with Differing Functions

Function (N)	Average Percentage to Progressive Farmers	Average Percentage to Noninnovative Farmers
General (88)	60	3
Coffee (10)	91	0
Animal husbandry (7)	57	0
Supervisory (13)	52	4
IDA loans (19)	39	5
Cotton (9)	57	19
Veterinary (32)	51	17
All (178)	57	6

than spreading hybrid corn to the present nonadopters. This set of priorities is difficult to justify by any criteria, as the marginal increase in output is usually greater with adoption of the new variety than it is from improvements in the quality of cultivation.

From the foregoing, it should be clear that the progressive farmer strategy does not provide the extensive distribution of benefits that its diffusion advocates have claimed. A broader range of extension contacts would probably lead to more rapid and widespread acceptance of profitable innovations. Further, that the bulk of these services is being provided to the progressive and wealthier farmers means that they also are helping to increase the gap between the rich and the poor. We do not mean here that rural inequality is created by the agricultural extension services. In a nonsocialist economy the farmers who are already somewhat better off than their neighbors are in the best position to invest in new, profitable farm enterprises, and we must expect that they will do so and hence increase their wealth.[19] If the farm economy is based on land, labor, capital, and knowledge, those who have more of these will make more money from their farming. But it does not follow that those who have the most of the first three should also be provided with a disproportionate advantage in technical knowledge by extension workers.

We believe that the middle group of farmers may be a more appropriate focus for extension than the progressives, for reasons of both equity and maximum diffusion. There is good reason to believe that poorer farmers will be quicker to adopt agricultural innovations from farmers who are basically like themselves than from the socially elite progressives. Although evidence is lacking, it also seems likely that profitable innovations will spread faster from less innovative to progressive farmers than they do in the other direction.[20] A highly innovative farmer with access to reasonable amounts of capital (and this is the definition of a progressive) will be quick to hear of profitable new products and techniques and will seek them out for himself from a neighbor. This self-drive does not characterize the middle or non-innovative farmers to anything like the same extent. When one is dealing with agricultural changes capable of general acceptance, we therefore suggest that adoption will be maximized in the long run by avoiding those farmers most anxious to innovate and concentrating on those who would normally be considered marginal.

But we have already pointed out that most agricultural innovations can be adopted by only a small proportion of the farming population. Furthermore, many of these new products and techniques require access to above-average amounts of land and capital. Wide acceptance is not a relevant criterion for assessing extension strategy on such innovations. Nonetheless, we believe that middle farmers are still the appropriate focus of attention for many of these change programs as well. There seem to be a significant number of middle farmers who have sufficient land and capital for certain innovations and who are passed over in the rush to their progressive neighbors, who have the most of these resources. Both a better distribution of wealth and a more specialized small farm economy will be gained if farmers who already have a major cash crop are bypassed in the extension of new ones.

The argument here for a focus on the middle farmer leaves the poor farmer, who lacks adequate capital, out in the cold. In view of the permanent and cumulative additions to rural inequality caused by the constant repetition of this innovation process, it needs to be seriously considered whether subsidized credit now may be preferable to public welfare later for the landless poor who are being created by it. At a minimum, campaigns on crops which have high labor but low land and capital requirements, such as vegetables, should be directed almost exclusively to the marginal farmers, who so desperately need them.[21]

We have shown serious deficiencies in the diffusion of innovations justifications offered for the progressive farmer strategy. The economic arguments for the progressive farmer strategy now must be faced. It is much more difficult to find logical fault with this set of justifications, for the inegalitarianism of the approach is openly accepted. "Betting on the strong" maximizes economic growth. Since the Kenya government acknowledges that its first concern is with growth,[22] it is legitimate to challenge its agricultural extension employees (rather than their minister) only if their distribution of services goes beyond what growth alone would justify.

It is useful here to examine a part of the extension services where the economic argument can be divorced from the diffusion one. Such a case is offered by the veterinary services. Visits by Animal Health personnel to individual farms are made almost exclusively for the treatment of cattle disease. Since less than 5 percent of the cattle in Western Province in 1970 were the economically highly prized grade variety, the cattle needing treatment may be considered as broadly equal in their innovation demonstration effect. The Ministry of Finance and Planning survey quoted earlier indicates that the progressive 10 percent of the Western Province farmers own approximately 16 percent of the cattle. Yet their farms receive 51 percent of the veterinary calls. On the other hand, noninnovative farmers own about 33 percent of all cattle and receive only 17 percent of the attention of the Animal Health personnel. Our very rough estimate is that each year 1 out of 55 of the cattle in the province receives a sick call from an Animal Health staff member (1.8 percent). Yet 1 out of 19 of the cattle owned by progressives will be seen (5.8 percent), while 1 out of 93 (1.1 percent) and 1 out of 111 (0.9 percent) of those owned by middle and noninnovative farmers, respectively, will be visited.[23] As can be seen from figure 9, veterinary services are distributed relatively equally between middle and noninnovative cattle owners. Progressive farmers, however, receive over five times better service than these other two categories. The top-to-bottom ratio of 6 : 1 would seem a vast improvement on the 44 : 1 observed for strictly agricultural services. But the latter has the diffusion of innovations theory as a rationale while the former does not.

Are there grounds of economic efficiency that might support this unequal distribution of veterinary services? A first response may be that progressives have larger herds and that a visit to one of their farms will be more efficient because of the greater number of cattle treated at one time. But this argument would be invalid, for there is actually very little difference in herd size between the categories of farmers for those actually owning cattle. Noninnovative cattle owners have an average of 5.9 head; middle farmers, 6.4; and progressive ones, 6.6. The overall inequalities in cattle wealth between these groups, which were discussed earlier in conjunction with table 27, were largely caused by the differing proportions of those with no cattle in each category.

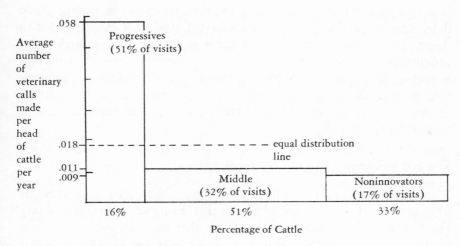

Fig. 9. The distribution of veterinary extension visits to farmers

Further justifications for the skewed distribution of veterinary calls might be offered. The progressive farmers probably take better care of their cattle, so that their cows will generally be producing more milk and have greater economic value. They also will be more likely to dispose of the economically relatively unproductive males, so that their herds will frequently be more valuable on these grounds as well. Finally, the better-off farmers are more likely to be willing to pay the small amounts of money necessary for medicines. Each of these arguments is basically valid and between them we can undoubtedly account for a portion of the differences in veterinary services provided to the three categories of farmers.

Nonetheless, the distribution in figure 9 has one feature which makes one reluctant to accept these economic explanations as adequate: while there is a substantial gap between the services provided to progressives and the other farmers, there is very little inequality in the distribution of visits between middle and noninnovative farmers. As middle farmers are situated between progressives and noninnovators in the modernity of their agricultural practices, it seems reasonable to expect that they would also be somewhere midway in their animal husbandry. If this were so, however, it would be economically efficient for them to receive fewer services than the progressives, but they should also be getting more than the noninnovators. But this is not the case. Thus we are led to suspect that there are additional, noneconomic grounds for the special emphasis on progressives. This conviction is further strengthened by

Njooro's study of the distribution of veterinary services among grade dairy cow owners in central Kenya. He found that Animal Health Assistants tend to respond more quickly to reports of sick cattle from progressive farmers than to ones from nonprogressives.[24] Although economic efficiency criteria could be used to justify a stress on progressive farmers in the distribution of both agricultural and veterinary services, the actual favorable allocation goes beyond anything that is economically rational.

The Social Causes of Unequal Distribution

What factors other than extension strategy might lead to a skew in services toward the progressives? In his Rungwe, Tanzania, study Thoden van Velzen offers the hypothesis that the unequal distribution of agricultural services arises from a social class alliance between extension staff and the wealthier farmers. He notes that Rungwe extension agents constitute a cohesive and somewhat isolated elite group and that those locals with whom they do have social contact are almost invariably rich farmers. He argues that extension staff have built up a symbiotic relationship with these rich peasants, involving the latter's providing land, food, and assistance on government projects to the staff. The staff in turn help the well-off farmers with access to government aid, support their dominance of local political institutions, and assist in their conflicts with other peasants. The consequences of this social class alliance are that the staff are reinforcing and accentuating inequality within rural Tanzanian society. Their isolation from poorer peasants is such that they seem to learn little from them and to provide them with relatively little in the way of direct positive benefits.[25]

Thoden van Velzen's analysis is provocative and important, and his hypothesis of a social class alliance between extension agents and rich farmers deserves a careful test in Kenya. As we saw in chapter 3, it is true that the ministry's senior staff belong to the national elite and that the junior staff fall approximately in the middle of the rural elite. Staff also associate very largely with the well-to-do in their social contacts with farmers. The approximately 90 percent of the rural population which falls into our Low Status category receives only 20 percent of the junior staff's friendship choices (see table 7). The contact of senior staff with the poorer farmers is even smaller (see table 8).

Is it then true that agricultural extension staff visit progressive farmers because they are their acquaintances or are the people most like themselves socially? Our evidence indicates that the answer is no. There is no positive correlation between the percentage of an agent's friends who are high and upper middle in status and the proportion of his visits devoted to progressive farmers ($r = -.12$, $p = .08$). If anything, there is a slight tendency for those who name the smallest percentage of friends in the high and upper middle groups to give a larger proportion of their time to progressive farmers than do the staff who identify more with the elite. Nor does it seem satisfactory to argue, as Thoden van Velzen does for Tanzania, that extension services are being provided to progressive farmers in Western Province as a reward for their help in official and private affairs. This exchange of benefits definitely occurs in Kenya, but it will not serve as a primary explanation for the stress on progressive farmers. As can be seen in table 28, those services most desirable to progressive farmers—loans and veterinary medicine—are better distributed among the three classes of farmers than are the other types of extension visits. We have already discussed the distribution of veterinary services. It is sufficient to add here that although progressives receive an average of 57 percent of all extension visits, only 39 percent of the loan investigations are made on their farms. If the coveted extension services were simply being given to those who would pay for them, the distribution of items such as loan investigations would show an even greater skew in favor of the wealthier farmers than do the less desired services. Since the opposite is the case, we conclude that pay-offs do not exert the major influence on the total pattern of inequality in extension visits.

Material collected by Solomon Njooro among a random sample of dairy farmers in central Kenya provides tentative evidence that bribery may actually even be reducing the extent of inequality in the distribution of veterinary services. First, Njooro asked the junior veterinary staff themselves (AHAs and JAHAs) to keep records on all farmers who reported sick cattle and on the cases actually treated by the staff. These records indicate that 20.2 percent of the reported cases received no services. Those whom the staff report to be members of the rural elite (prominent leaders and those in wage employment) had their requests for service unmet only 6.7 percent of the time, where the nonelite went without assistance in 27.9 percent of the cases (see table 29).

Njooro also questioned a random sample of farmers about the response they had received to their requests for veterinary calls. Among those who were actually able to get service, a much larger proportion of the nonelite than of the elite alleged that they had had to pay for the

TABLE 29
The Response of Veterinary Staff in Mathira Division,
Central Province, to Sick Cattle Reports

| | Request for Service | | |
Status of Farmer	Met	Unmet	Both
Elite	58 (93.5%)	4 (6.5%)	62 (100%)
Nonelite	80 (72.1%)	31 (27.9%)	111 (100%)
Total	138 (79.8%)	35 (20.2%)	173 (100%)
	$\chi^2 = 11.45$	$p < .001$	

SOURCE: Solomon N. Njooro, "Distribution of Veterinary Services in Mathira Division of Nyeri District," p. 29. Drawn from records kept by AHAs and JAHAs in June 1972.

supposedly free call. These results must be interpreted cautiously, as the sample is small and the observed differences are not statistically significant (see table 30). Nonetheless, it is certain that members of the elite do not make extralegal payments for the veterinary services they receive *more* often than the nonelite. There is a reasonable possibility that they have to pay less often. Thus it would appear that the rural elites have established a nearly automatic right to extension services. One of the ways open to the nonelite to overcome this skew in assistance toward the well-to-do is to pay for it. That loans and veterinary calls are highly valued and therefore worthy of a bribe may well explain why they are better distributed among the different status groups of farmers than are other extension services. In a system that is already heavily oriented toward the elite, bribery may actually mitigate inequality in the distribution of benefits.

The idea of a local social class alliance is not sustained by our evidence and the hypothesis of an exchange of benefits is able to explain only a small part of the excessively large skew in services toward progressive farmers. What factors then are at work in causing even more inequality than the ministry's extension policies would dictate? The most important reason is probably the squawk factor—that the progressive farmers are the ones most likely to complain to a senior officer if extension is not provided to them. We have already seen that junior staff do only a small amount of work and sometimes organize themselves informally to reduce the amount of effort they put into their jobs. As the work of visiting farmers is carried out in a very wide area, the junior staff are largely free of any supervision. Complaints are one of the very few ways that an Assistant Agricultural Officer has of judging

TABLE 30
Extralegal Payments for Veterinary Services in Mathira Division, Central Province

Status of Farmer	Received Service		
	Paid for	Not Paid for	Either
Elite	4 (22.2%)	14 (77.8%)	18 (100%)
Nonelite	4 (66.7%)	2 (33.3%)	6 (100%)
Total	8 (33.3%)	16 (66.7%)	24 (100%)
	χ^2_y = 2.25	$p < .20$	

SOURCE: Njooro, "Distribution of Veterinary Services in Mathira Division of Nyeri District," p. 30. Drawn from interviews with a random sample of dairy farmers in June 1972.

whether his subordinates are on the job. Thus the rational extension strategy for the agent who wishes to minimize his effort is to see the complainers and forget the rest. By virtue of their relative wealth and their past innovativeness, the progressive farmers are among the few who will have the self-confidence actually to complain to an officer. Thus the progressives do have a power advantage. But it is not one born of a local social class alliance, exchange of benefits, or direct political influence. Their power often derives quite simply from their presumption that government should provide them with services and that they can tell an officer if it does not. The general structure of class privilege in the society does interact with the client's social background, education, and occupation to affect his expectations of service and his manner in demanding it. But the extension agent is primarily responding to the character of the demand, not the social class of the client as such.

It may be an unfamiliar idea that a client of a bureaucracy should be more powerful in dealing with it simply by believing that he has a right to its services and by being willing to complain if he does not get them. Three incidents observed by Solomon Njooro in Mathira Division may help to illustrate the social forces we have in mind. In the first, an Animal Health Assistant (AHA) was sitting under a tree at the chief's office waiting for veterinary calls. Before noon an older farmer came up to him hesitatingly and told him that his grade cow was sick and needed attention. The AHA did nothing and finally the petitioner retired to the other side of the clearing and sat down to wait. Nothing more was said; no one else came; at 4:30 P.M. the AHA announced that his work day was over, got up, and left; with resignation the old farmer did likewise. On another occasion Njooro saw a part-time farmer with wage employment approach the same AHA under the same conditions. This time, however, the petitioner was self-assured and clearly expected to be helped. The AHA responded immediately.

In the third incident, a middle-aged woman came to the house of the divisional Veterinary Officer about 5:00 P.M. to demand that he come to see her cow, which was having major problems in calving. VOs are under no official obligation at all to service clients after working hours, although they can and do take such cases on a private payment basis. The Mathira VO said as much, but the woman demanded her "rights" to free help and refused to take no for an answer. For the next two hours she stayed at the VO's house pressing her case loudly and ceaselessly. She even telephoned the Provincial Veterinary Officer to complain, but he supported the Mathira VO. Shortly after 7:00 P.M. the VO finally gave up, took a government vehicle, and went to help the woman. In no one of these cases was the client known personally to the staff member, nor was a bribe asked for or offered. The differences in service provided were due directly to the nature of the clients' demands.

Our hypothesis that extension services are skewed toward potential complainers is strengthened by two sets of facts. One is the slight tendency, reported earlier, for the lower status agents to give a greater proportion of their visits to progressives. These junior staff are less secure in their positions and so must devote more care to preventing complaints. Similarly, the best-educated extension workers make the smallest number of visits all together (see chapter 6). They are more secure and better able to survive unfavorable reports to their superiors.

The other set of facts supporting our squawk factor hypothesis is that the less extension supervisors rely upon complaints being brought to them and the more they get out in the field themselves to learn actively about the work of their staff, the better the distribution of their subordinates' services. The more often the first and second line supervisors (LAAs and AAOs) interact with their junior staff, the smaller the proportion of visits received by progressives ($r = -.68$, $N = 14$, $p < .01$; $r = -.48$, $N = 14$, $p < .05$). Similarly, it appears that the skew toward progressives may tend to decrease as the two superiors (LAA and AAO) put more effort into supervision ($r = -.41$, $N = 14$, $p > .05$; $r = -.34$, $N = 14$, $p > .05$).

Extension for the complainers is buttressed by the distorted picture that government often has of the small-agriculture world. Senior staff in particular are likely to have an optimistic view of the degree of acceptance of modern farming. Joseph Ascroft was told by agricultural officers in Nyeri that Tetu Division had 100 percent acceptance of hybrid corn, but his random survey of 354 farmers found only 31 percent growing the crop.[26] The social isolation of senior staff from the areas in which they are working helps sustain these distorted perceptions. Even junior staff, who are drawn largely from the communities in which they work, have an optimistically biased view of their areas. This is well illustrated by the reactions of the AAs who conducted the preliminary survey of farmers for the current Vihiga Division extension experiment. Confronted with a genuinely random sample, they confessed that they had never realized that such poor people even existed in the areas in which they were working.[27] It is much easier to ignore the nonprogressives if one is unaware of their proportionate importance.

Attention to the complainers and the invisibility of the rural poor probably account for a substantial part of the inequality in the distribution of extension. Nonetheless, we do not want to forget that the dominant explanatory variable is the ministry's acceptance of the progressive farmer strategy and its supporting ideas of the diffusion of innovations and the maximizing of economic efficiency and growth. We have accepted that the economic growth arguments are logically sound (even though we doubt their morality and political wisdom). Certainly an emphasis on growth rather than equality needs no special explanation in Kenya's political system. We have noted, nonetheless, that the actual unequal skew in extension seems to go beyond what could be justified on grounds of economic growth alone.

More interesting is the acceptance of the diffusion of innovations idea when its assumptions do not seem to fit Kenya's economic and social realities. The strength of this belief system must lie in good part in its having been dominant in almost all of the agricultural education institutions (in Kenya and abroad) in which extension personnel have been trained and socialized. As Blau and Scott have noted, it is often the marginal members of a social group who show the strongest and least discriminating support for its dominant values.[28] The lower status members of junior staff feel insecure in their position and they note that their supervisors voice a belief in working with progressive farmers. As a result they may skew their services toward progressives beyond what their superiors, who understand the logical limits of the strategy, would do themselves.

There is in addition probably an unconscious mechanism supporting the belief system—a visit to a progressive farmer is simply more satisfying. One can expect to encounter less resistance to new or difficult farming practices, and one is more likely to see a change on that particular farm. Therefore, the agent feels he is getting better results, as did the extension workers polled in Tanzania by Saylor. It is emotionally difficult to accept that a better long-run, total impact may be achieved in one's area by working with somewhat less receptive farmers. Net effects are hard to see, but the contacted farmer is immediate and real.

Of course it is far from accidental that so many different factors combine to produce inequality in extension service. There is a larger network of class politics working to sustain most of them. British colonial extension policy in the 1950s and Western extension thinking in the 1960s were explicitly focused on progressive farmers in order to create a rural class congenial to capitalist development.[29] Thus almost all extension personnel throughout the world consciously or unconsciously understand their work within that framework. Although leading figures in Western development circles, such as at the World Bank, now see capitalist growth as better served by extension for the poor, it will take a long time for this change in ideology to alter the training of the men in the field, even in socialist-leaning countries.

Similarly, Kenyan policymakers use economic growth rationales for their programs because in doing so they legitimate benefits for the African middle class groups to whose interests they are committed.[30] The same forces permit the continuation of an emphasis on progressive farmers beyond what can be justified by growth and diffusion of innovation possibilities. Even the squawk factor is ultimately rooted in the class structure of Kenya. Wealthy and progressive farmers believe that they have a right to demand extension services because the social and political structure gives them privileged access to everything else. Thus scholars such as Thoden van Velzen are correct that class analysis illuminates the structure of service distribution in agricultural extension. They are wrong, however, in seeing these inequalities as deriving mainly from a class alliance between local elites and junior government staff. The latter are really only tools of larger, systemic pressures at the national and international levels and are manipulated by the social expectations, training, policies, and organizational pressures that we have detailed. There are forces at the local level for inequality in service distribution, but these are much less important than the national ones. Changes at higher levels could easily lead to much greater equality in the distribution of extension benefits, even in conjunction with an emphasis on economic rationality and efficiency.

Conclusions

There is a substantial bias in the distribution of agricultural extension services in Western Kenya in favor of the wealthier and more progressive farmers. This favoritism accentuates rural inequality and probably prevents the maximum possible acceptance of agricultural innovations. The bias of junior staff toward wealthier farmers seems to be best explained by the progressive farmer strategy, even though part of its diffusion of innovations ideological underpinnings are deficient and the skew goes beyond any emphasis that could be justified by economic growth arguments alone. The squawk factor is the other major variable explaining the skew and comes from a combination of a weak commitment of junior staff to their work, the pattern of farmer demand for extension services, and a somewhat distorted perception by agents of the proportion of rural societies made up of progressive farmers. An even more optimistic view of their areas is held by the senior staff and is doubtless sustained by the isolation of all except the officers of local origin from their subordinates and the local community. The major factors we have advanced as leading to the disadvantage of the less wealthy farmers are rooted in the past and present history of class politics in Kenya, mainly at the national and international levels, not at the local one. Nonetheless, they are organizationally manipulable, given the necessary political will among Ministry of Agriculture decision makers. The skew might well be lessened by Ministry of Agriculture programs that carefully redefined extension strategy, developed specific guidelines for working with the middle or even bottom rungs of farmers, and gave the agent some solid basis for resisting progressive farmer demands. One possible way of doing this would be to define extension goals by numbers of cultivators moved over a defined threshold of farming quality. The organizational pressures to work with those above the mark would then be removed.[31]

This is not to say that the skew could be completely eliminated by organizational (or any other) devices. Ascroft and his colleagues drew up careful and specific guidelines for subchiefs and extension agents to get them to recruit noninnovators for a special farming course in Tetu (central Kenya). They succeeded in getting a group of which a third were middle farmers and a fifth noninnovators. This was a substantial improvement on past inability to reach any of these groups of farmers. But half of those sent by the field staff were progressive farmers who had not previously attended this type of course. The Ascroft team credit the inclusion of the progressives despite explicit instructions to the contrary to the field staff's not knowing the characteristics of the farms that they do not visit. The agents did not know that some of the farmers selected actually were progressives and, through past neglect, they did not even know who many of the noninnovators were.

If ignorance were the sole problem, programs that repeatedly forced the field staff to work with the poorer farmers would help to overcome it. An additional factor, however, is that those who attended this particular course were to receive special credit for farm inputs. It is likely that the desire of the field staff to send only worthy farmers or to earn favor with certain formerly ineligible candidates was also working for the inclusion of progressives.[32] Inequality in the distribution of services is a tendency with strong roots in social behavior, and it is unlikely that it can ever be fully eliminated. At the moment, however, the skew toward progressive farmers by the Kenyan extension services is being substantially accentuated by policies and organizational structures that are not even necessary or rational in Kenya's overall approach to rural development.

The contributions of the Ministry of Agriculture to rural inequality can and should be eliminated. Whether they will be is open to more question, however. Although the skew in services does not even make good economic sense, it is consistent with the general pattern of benefit distribution in the Kenyan political system. Decision makers probably will not feel any urgency about ameliorating the imbalance in extension until poor farmers emerge as a distinct and noticeable political force and are able to campaign for their own rights.

Part 4   Conclusions:
                Practical
                and
                Theoretical

10 Organizational Planning and the Depersonalization of Structures

The Colonial Legacy

Agricultural extension administration in developing countries is organization-intensive. In addition to arranging for farmers to be informed of technical innovations and to be taught good agricultural practices, it is involved in assuring that supplies and credit are at hand when needed and that markets are readily available when the products it is encouraging are ready. When these responsibilities are added to the unpredictability of the weather and the absolute deadlines that characterize planting and harvesting cycles, the administration of agricultural extension becomes particularly complex. In this chapter, we examine the relationship between an organization's structure and its ability to meet the challenge of extension administration. In so doing we draw together into a coherent set of recommendations the many conclusions of the preceding chapters. The insights that we have gained on worker participation, styles of supervision, motivation, communications, feedback and inequality all have implications for the way in which an extension service is organized. As is the case with all practical advice, the details of our proposals are specific to one country—Kenya. Nonetheless, the basic features of our recommendations are applicable to most developing countries and will provide a useful starting point for discussions of administrative reform in their extension services.

Jon Moris provides a useful perspective on issues about overall organization structure in a pair of essays on the history of agricultural extension administration in the Embu District of Kenya.[1] Moris points out that the Ministry of Agriculture requires a large degree of "organizational intelligence" if it is to be able to deal effectively with the complexities of its tasks. This intelligence will exhibit itself in the ministry's ability to anticipate opportunities and problems, to plan wisely to meet them, and to control the actions of its own staff in order to implement its decisions effectively. The word *planning* may imply that organizational intelligence is a headquarters function. This need not be the case. The identification of and solution to problems could as well be left to widely dispersed field officers. Different methods of structuring the total organization may be used to try to achieve a high degree of organizational intelligence.

We can refer to any organization as having a set of design specifications, which are its basic structural and behavioral features. Among the design options that Moris mentions as being open to field extension organizations are the following:

1. The type of contact group it tries to build up . . . [in] the local community.
2. The professionalism and training of field workers at each level.
3. The nature of supervision, sanctions used, and spans of control between levels within the agency in the field.
4. The length of downwards communications' chains, and the provisions made for upwards communication.[2]

Moris does not deal with all of these issues in design specifications, though we shall. He does offer a useful analysis of the field control structure inherited by the ministry from its colonial predecessor.

When an organization is first being created, there may be careful consideration of what its design specifications should be. Once set, however, "they are usually subsumed into the organization's inventory of 'routine procedures.' As such they rapidly become innocuous and invisible. . . ."[3] This is precisely what has happened to the colonial system of field organization, which has been taken over intact by much of the Kenya government.

The colonial management pattern was to give great independence, discretion, and responsibility to all European field officers and virtually none to their African subordinates. Moris has termed this the hub-and-wheel pattern of authority. He argues that this form of managerial control

tends to emerge naturally in any organizational setting where large numbers of lowly qualified staff are working under the supervision of a few key professionals. Told to organize complex activities from scratch without supporting professional staff, the centre-post man sets up a circle of subordinate workers each capable of doing one limited aspect of the total operation. Thereafter, it is always quicker for the centre-post man to perform the difficult tasks himself than it is to train someone else each time a special skill is required. The subordinates soon become accustomed to relying on the centre-post man for all inputs of a supervisory, planning, co-ordinating, or decision-making nature.[4]

Moris is correct that the hub-and-wheel pattern was a rational response to the situation in which the early colonial Agricultural Officers found themselves in the 1920s and 1930s. They were virtually on their own in devising, funding, and executing strategies for dealing with fairly complex problems of peasant agriculture. The staff assistance available to these AOs was unqualified and had to be completely trained and supervised by them.

The hub-and-wheel pattern also was probably psychologically satisfying to most colonial officers. As both Robert Chambers and J. Gus Liebenow have remarked, great independence in decision making and highly personalized programs were hallmarks of British colonial administration.[5] Many colonial officers had left Great Britain in order simultaneously to escape the confines of organizational control carried by industrialization and to gain personal rulership in another setting. The hub-and-wheel system gave the officer the independence he desired. It also kept his African subordinates in a completely inferior position by not incorporating them into the planning or control functions of the organization. The officer's need for rulership was thus met in a way that fitted his conscious and unconscious ideas of racial superiority. That the hub-and-wheel pattern was psychologically satisfying to the European officer must account in good part for the lack of energy with which the colonial Department of Agriculture met the problems of training and encouraging African professional and paraprofessional staff.[6]

The African officers who had been socialized to the hub-and-wheel pattern of authority from beneath in the colonial period kept it in their organizations' design specifications when they succeeded to positions of responsibility. Despite the almost total Africanization of the Kenyan civil service, the behavioral patterns of its members still betray those who hold "European" positions and those who are to be treated like "Africans."

Yet the personalized hub-and-wheel pattern had ceased to be efficient by the end of the colonial period and today it is even less suitable. The conditions which first made the pattern rational no longer exist. African junior staff are no longer untrained; they have gained paraprofessional status. Furthermore, the price of their labor is no longer so low that it can be used inefficiently. Similarly, the preconditions of the Agricultural Officer's independence have ceased to be met. There are now a sufficiently large number of professional officers in the field that their efforts do overlap and need coordination; development programs in African small-farming areas are no longer initiated and executed without central funds and thus without the need for approval and cooperation from headquarters officials. The result is a system of hierarchical control neither loose enough to permit a civil servant to fulfill his present responsibilities on his own nor tight enough to force him to do so. Moris gives the example of a cotton extension campaign in Embu where an officer can be seen to be independent enough to fail to take the actions implicitly expected of him but too dependent on cooperation from his fellow field officers and Nairobi to complete a development initiative successfully on his own.[7]

In addition to the effects of change in the organizational environment, the younger African officers do not have the driving authoritarian character of their European predecessors, for whom the hub-and-wheel pattern was tailormade. We interviewed thirty-four current agricultural extension agents who had held the same position in the colonial period. When asked to describe "the differences between the old European AAOs and the present African ones," 56 percent (19) volunteered that the Europeans were particularly bad in using force and "making noise." The result was that the junior staff worked harder than they do now.[8] Several of them point out, however, that this did not necessarily mean that they had accomplished more. The European AAOs generally had less of an idea about how hard or how well their subordinates were working than the present African AAOs do.[9] As was true of other colonial bureaucracies, junior staff often were working hard at maintaining the appearance of activity.[10] The compulsive force that characterized colonial administration and made the hub-and-wheel system work is now found only among the older, colonially trained officers.

It seems to us that the young African officers in the ministry do not have the same deep drive for personalized achievement and individual control as did their predecessors. They are more team men than authoritarian individualists. Others of their virtues are different as well. Eight of the junior staff whom we interviewed volunteered that the African AAOs are much better teachers than the European ones and five mentioned that they are less distant and in better communication with their subordinates. The hub-and-wheel pattern requires the officer to be the personal source of overpowering authority. The present AAOs would have their strengths better utilized by a system of institutionalized and routinized authority.

Supervisory Systems and Group Extension

Many of the shortcomings in Kenya's extension services can be traced to a chronic overloading of its senior staff field officers and to the inadequacy of the organizational structures that they have to work with. Let us reverse the usual order and begin our examination of structural problems at the bottom of the extension hierarchy. Junior staff agents do half the work that could reasonably be expected of them and the divisional AAOs who supervise them are powerless to overcome their collective resistance. The felt need of the divisional AAOs to be personally authoritarian with their subordinates in order to get them to work causes them to be very poor at stimulating the upward flow of information. Consequently, the innovative and adaptive functions with which they and the district Agricultural Officers are entrusted are poorly fulfilled. AAOs are also so busy chasing after their junior staff that they have little time to anticipate and solve supply and marketing problems in their areas. Furthermore, AAOs are expected to keep their agents well-informed, but the ministry's information system actually operates in a way that makes them sources of authoritative misinformation. Divisional and district senior staff are sufficiently tied down with trying to make an inadequate administrative system work that they do not have the time to perform the tasks for which their professional training best suits them—the analysis of technical agricultural difficulties.[11]

The overload on the senior staff field officers could be greatly reduced by a management system that enabled the routinization and delegation of many of their supervisory functions. We have remarked several times on the extreme difficulty of checking on the performance of subordinates when they are working over a wide geographical area. The LAAs supervise a mean of nine extension agents and the divisional AAOs have on average thirty-five junior staff under them. As an agent can be working legitimately anywhere within his area of responsibility, it is difficult to locate him without advance notice and therefore almost impossible to verify that he is on the job. The supervisory dilemma posed by geographical dispersion is exacerbated because most LAAs basically identify with those whom they supervise and can be counted on to report only the gravest delinquencies.

In the circumstances, the major burden of ensuring that junior staff are at work falls upon the divisional AAOs, who are ill-equipped to deal with it. Most rely on complaints from farmers to identify the lazy agents. This results in the disproportionate attention given by junior staff to progressive farmers, who are the most self-assured of the potential malcontents. The other supervisory technique employed is to require extension agents to wear uniforms on the job and then to keep an eye out for uniformed men loitering in the bars and shops.[12] Both of these supervisory methods note subordinate performance only in default. Neither technique identifies hard workers; both pick out only the worst offenders. When an AAO wishes to form a more complete opinion on the work of his staff, he has to spend long hours traveling around the countryside, looking for agents and checking with farmers. Even then he will be doing more fault-finding than accomplishment-praising. Under present conditions, if a divisional AAO attempts to keep his subordinates on the job, he has to spend a great deal of time at it and exercise his authority in a personal and unpleasant manner. The time demands of supervision keep him from the tasks for which his professional qualifications suit him. An appearance of personal authoritarianism undercuts staff commitment still further and hurts the possibilities for feedback from his subordinates.

To the best of our knowledge the Kenya Tea Development Authority (KTDA) is the only extension organization in East Africa that has made a significant attempt to solve this supervisory problem. The KTDA has registered all tea growers and given each extension agent responsibility for a specific set of these farmers. The agent is required to record each farm visit in a book, which is collected semiannually and out of which statistics on the number of visits made by each agent are computed. The use of explicit job targets and of work records represents a major advance on the anarchic system employed in the general extension services and is generally believed to have increased the productivity of the tea agents.

Nonetheless, the KTDA system does have three problems. First, the agent's self-recorded work records need to be verified if the system is to be fully effective. Although in principle this task could be done through supervisory visits to a systematic sample of farms for each agent, such substantiation is not actually done. It would be too demanding of the expensive time of the Tea Officer. The amount of time involved in traveling between farms and the essentially negative character of a check on the reliability of records also make the AAs in charge of the local groups of agents shy away from meaningful verification. As a result, the work records are in fact a carefully kept fiction.[13]

Second, the system depends on the keeping of a comprehensive register of farmers and the characteristics of their farms. Even the KTDA register of tea growers has many erroneous entries.[14] If the register system were extended to all farmers and not just the few with tea, it would become so burdensome as to be unsustainable. In fact, colonial attempts at farm registers all collapsed after a few years of strenuous effort.[15]

The third problem with the KTDA method of supervising staff would arise if the first two were solved. The quantity of extension visits is one aspect of an agent's performance but the quality of advice and services offered is also of vital importance. It is a well established fact of organizational life that when workers are judged on the basis of performance statistics, their efforts are directed toward the thing measured and away from that which is not.[16] Thus there is a great danger in formalizing and emphasizing visit records like those of the KTDA, which are only partial measures of performance.

We feel that a shift to group extension methods would provide the kind of framework within which supervisory control and planning would become feasible and would produce a much more effective extension service. Group methods have been finding increasing favor with agricultural extension services in the developing world.[17] Larger numbers of farmers can be reached; method and result demonstrations can be provided more economically; and there is considerable evidence that it is easier to persuade people to adopt innovations when they are in groups than when they are encountered as individuals.[18] Although the bulk of Kenyan agricultural extension services is provided through visits to individual farmers, a movement toward group methods, especially for demonstrations, is underway. Schönherr and Mbugua conducted a very successful experiment with group extension in Nyanza Province, Kenya, during 1974. Eight extension agents worked with an average of 11.8 farmers each, organized into a group. All of the 94 farmers involved both planted and marketed the new crop being extended (soya beans). In an experimental control, three agents worked with four farmers each as individuals. All the twelve farmers planted the soya beans but only seven (58 percent) followed through with the crop and marketed it. Thus the group extension methods both achieved a greater proportion of effective adoptions and reached a larger number of farmers per agent. Furthermore, the farmers taught in groups achieved dramatically higher yields than did those who received personal instruction. Subsequent trials with which Schönherr and Mbugua have assisted are equally promising.[19] Group extension is clearly the best technique available for communicating technical information to farmers.

The process of creating extension groups can be illustrated from Migori Division in South Nyanza where the senior agricultural staff instituted a two-tier organizational structure. Agricultural Village Committees were formed for each sublocation, the area served by a generalist junior staff agent. An open public meeting was called in which committee members were elected from all the areas of the sublocation. The committee members in turn formed voluntary area-based groups in their neighborhoods. The committee members serve as chairmen of their groups and act as a liaison with the local extension agent. The Agricultural Village Committees make decisions on local agricultural matters and watch over the performance of extension staff and the standards of farmer cultivation.[20] Such a structure conforms to the suggestions of Norman Uphoff and Milton Esman for local farmer organizations. In a review of the rural development experience of sixteen Asian states, they conclude that

> The more successful systems . . . function with several levels of organization, most commonly as two-tier structures, though three or four tiers may be practicable. The primary level of organization tends to be the hamlet, the neighborhood or some other group of 30 to 100 families.[21]

Such groupings have sufficient social cohesion and organizational resources not only to be effective for extension but also to provide the eventual framework for needed cooperative ventures and shared services.

Group extension involves much more than convenient collections of farmers. In the Nyanza Province experiments, group leaders even came to perform extension functions themselves, although this is not essential to the method. Mbugua and his associates report that

> Communication between the extension staff and the farmers is conducted through the group's representatives and/or the group of farmers in toto. For example, the instructor might arrange the date for a planting demonstration with the group chairman. The chairman in turn communicates this message to his farmers, selects a demon-

stration plot and asks all farmers to come at the appointed time. The instructor then carries out the demonstration, and afterwards the group leader supervises implementation of the demonstration on the plots of the group members.[22]

The social cohesion of a neighborhood group makes the conclusions of its collective deliberations far more persuasive and the oversight of its leader more effective than anything a civil servant could provide.

Group methods are the best available in extension because they use the social dynamics of the farm community and because they employ demonstrations. They also have features that make them ideal for supervisory purposes. For example, a demonstration represents the culmination of a long period of agent preparation. Result demonstrations have a particularly long gestation period: a farmer must be found to try the crop or innovation; the appropriate materials have to be made available; every stage in the progress of the crop needs to be explained and supervised; input records should be kept; and neighboring farmers must be persuaded to attend the meeting at which the results are to be assessed and discussed. Method demonstrations and lectures are less onerous but still involve a significant amount of preparation. An agent or local group leader probably needs to pay quick visits to a hundred or more farms in order to generate reasonable attendance at any kind of meeting, particularly in the early stages of his organizational work in a community. A good demonstration will often generate demands for follow-up visits or assistance in obtaining supplies, putting still more work on the agent. Although we do not have the information necessary for a sound judgment, an agent probably cannot average more than two meaningful demonstrations per week of work.

From the supervisory point of view, the advantage of demonstrations and group meetings is that they represent an inspectable, relatively infrequent, final product of a considerable amount of extension work and the observation of them is likely to tell one almost everything one would like to know about one's junior's performance. All kinds of things can be ascertained from one hour at the demonstration or lecture of a subordinate. Technical competence is evident in the presentation and demonstration of information; preparatory publicity work can be gauged by the attendance; the extent to which the agent is concentrating on the right farmers can be learned by asking farmers quick questions about themselves; the amount of effort going into follow-up work and provision of supplies is easily estimated by asking those present about their past receipt of these services. In fact, the meeting is a comprehensive enough activity that its various aspects could be used as the statistical basis for evaluating agent performance without creating an unfavorable distortion in his work effort. The use of meetings and demonstrations for agent ranking is facilitated by their being given at fixed places and times, known well in advance, so they can be visited unannounced by a supervisor. Thus an LAA who had a motorbike and spent only half of his time on supervision could easily inspect a quarter of the demonstrations given by ten subordinates. A divisional AAO who spent only an eighth of his time in field inspections could see two demonstrations a year for each of a staff of forty subordinates. In this way an AAO could validate the evaluations of junior staff being made by the LAAs without too great an input of his own time. Reliance on the group meeting and demonstrations therefore makes extension agent work truly visible and renders possible for the first time the meaningful evaluation and reward of worker effort.

Beyond their advantages in reaching farmers and as a supervisory method, group extension offers other substantial benefits. One is that it can reduce the skew in extension services toward progressive farmers. First of all, it is possible for the AAO to specify the types of farmers that he would like to see his agents concentrate on and then check to see if these are actually being included in the group meetings. Second, as long as the groups are based on relatively small areas—say an administrative village—and are required to be of medium size—say thirty to fifty members with varying attendance—they will include of necessity almost all middle income farmers and many poorer ones.[23] Access to technical information would then be based on willingness to attend meetings rather than on the farmer's confidence in demanding services. Groups provide the institutional framework that is currently lacking for farmers to share experiences with one another. Such group meetings would therefore help to overcome one of the problems with the diffusion of innovations from innovators to poorer farmers.

Group methods offer an additional managerial advantage in that they make overall extension program planning more feasible. As the actual use by junior staff of their work time is very nebulous at the moment, it is quite easy for an AAO to plan extension campaigns that will conflict with one another in their use of personnel without his being aware that he has reached this point of competition. For example, in one district, hierarchical demands on the labor of junior staff over a year's period varied from 18 to 474 percent of the man-days available.[24] If an AAO is dealing with fixed units of work that have known and easily blocked time demands, program planning becomes a meaningful act of choice for him. For example, if a comprehensive campaign involves conducting demonstrations in the area of each headman, agents are responsible for five such villages each,[25] and they can do two meaningful demonstrations a week, it is clearly impossible to conduct comprehensive campaigns for both corn and peanut planting in the four weeks of March. One or the other campaign will need to be slightly postponed or dropped, or they will have to be conducted for different areas. Under the system of planning with demonstration units, the conflicts are readily identified and the exact time and nature of the staff training and supplies needed for programs are evident.[26] As we noted earlier, once groups have been constituted and institutionalized, they also can become intermediate organizational units for use in the distribution of supplies and the transport of goods to markets. Such groupings would therefore help the AAO in the solution of other of his major problems.

Let us return, however, to the issue of using the evaluation of group extension meetings as a method of control. To be most effective, such evaluations would need to be made routine and reasonably objective. AAOs and LAAs would need to be required to make a minimum number of random checks on each subordinate's meetings. The LAA, rather than the AAO, would write the annual staff evaluation reports and the function of the AAO's checks would be solely to validate the objectivity of his LAAs' judgments. LAAs would be required to base a large part of their annual staff evaluation reports on their observation of the meetings and to record certain objective facts about them in their reports. For example, they could be made to state the average number in attendance at an agent's meetings, the number of follow-up visit requests that the meeting generated, whether any misinformation was disseminated in the meeting, and so on.

It may seem burdensome and restrictive to make staff evaluations routine and largely objective, but substantial benefits would accrue with these seeming impositions. First, the reduction of the evaluation process to procedures would make it possible to delegate it to the LAAs with confidence, relieving the AAO of his overload. Second, when a superior has the discretion to report or not report subordinate shortcomings or to report only an edited version of them, the decision takes on a personal tone and the supervisory relationship becomes charged with emotion. Subordinates who are rewarded turn into sycophants; those who are punished become aggressive or avoid all contact.[27] When, on the other hand, superior evaluation of subordinates is routine, required, and relatively objective, the supervisory relationship is more depersonalized; the superior is seen to have no choice but to report both outstanding and deficient performance. Paradoxically, when supervision is depersonalized through the use of fairly objective performance records, the personal relationships across hierarchical levels greatly improve. Rather than hiding their problems from their superiors, subordinates are likely actually to point them out and ask for help in their solution. The general tendency for impersonal mechanisms of control, such as performance records, to substitute for personal supervision, to reverse the flow of demand in the organization, and to improve relations between superiors and subordinates has been well documented by Peter Blau and Richard Scott.[28] Precisely the same advantages to routine, objective performance records have been demonstrated in Kenyan field experience with plan implementation reporting procedures.[29] Since one of the things that is needed in the Ministry of Agriculture is more and faster feedback concerning problems experienced at the junior staff level, there is every reason to favor the use of impersonal mechanisms of control over its extension agents.

Objectivity in staff evaluation should not become an end in itself, of course. If the procedural requirements for recording performance statistics become too narrow and rigid, an officer may be forced to recommend for promotion a subordinate who works only on what is measured, to the detriment of his full function, or to pass over a staff member who always volunteers for the most difficult tasks, giving himself a poor formal record. Both the supervisor and the supervised should have the opportunity to add their written explanations of the objective data to the formal record. But the performance statistics should provide the solid core around which all other evaluations must be organized. Reliance on routine and objective performance records also need not, and indeed should not, preclude consulting junior staff when setting their work targets. We have stressed that extension agents are by no means helpless in the face of authority and that through their work groups they are capable of defeating the wishes of their superiors. In the geographically dispersed setting of agricultural extension work, any AAO who challenges his junior staff to a contest of wills is sure to be the loser. It has long been the thesis of the human relations school of management that the best way to beat the work group is not to attack it but to seek its cooperation through participation.[30] Just as popular opinion has doubted the negative effects in Kenya of authoritarian leadership, many believe that subordinate participation is unnecessary to organizational effectiveness in Kenyan conditions. The evidence that we have at hand suggests that the second view is as mistaken as the first. Studies done in Kenya on both secondary school teachers and on divisional development project teams indicate that staff who are not consulted show less commitment to their work and much more resistance to their superiors.[31]

Furthermore, there is good reason to believe that when junior extension staff are asked to participate in setting their work targets, they will set responsible ones. We noted in chapter 4 that agents give a public definition of a fair day's work that is actually reasonable. The concrete experience of Chambers and Belshaw in letting junior staff participate in setting work targets in Mbere Division, Embu District, is that they do in fact accept levels of work which their superiors regard as fair.[32] The agents in work groups are in a weak position when they are actually consulted, for they themselves believe that they are presently doing too little work. If they are asked to participate in setting their targets, they will in fact accept reasonable ones, their ability to organize against these levels of work will be compromised, and their motivation on their jobs will be substantially improved.

Thus the ideal combination for supervisory purposes is: (1) easily evaluated task units, (2) work targets set in consultation among AAOs, LAAs, and junior staff, and (3) work performance evaluated regularly, routinely, and fairly objectively by the LAA, with (4) the AAO making validity checks on the evaluation reports. Finally, (5) these evaluation reports would need to become the basis for promotion in a meaningful career system, such as we advocated in chapter 6. We believe that such a set of reforms would substantially increase the productivity of Kenya's extension agents while simultaneously relieving the AAOs of some of their present fruitless supervisory tasks. Furthermore, although the increase in explicit procedures would make the ministry more bureaucratic in one respect, it would make the extension services less bureaucratic in another by enabling AAOs to become less authoritarian and more democratic with their subordinates. Not only would this be more satisfying to the junior staff, but it would also be beneficial to the African AAOs, who have good nonauthoritarian qualities that are now underutilized.

Linking the District and the Capital

The key concept in rural administration is *linkages*. The days of autonomous agricultural development are ended. The development of the small-farm sector is critically dependent on government support. The state develops new agricultural technologies, promotes and finances their acceptance, and determines the dynamism of their growth through price and other controls. Agricultural development autonomous of government does occur, but relative to the growth needs of today's poor nations, its pace is too slow. If laissez-faire is dead, however, central command agriculture has never been born. Just as the small farmer depends on the assistance of government, national ministries need local initiative if rapid development is to occur. There are too many unique local opportunities that are profitable to exploit and location-specific problems that need urgent solution for rural development to be commanded completely by central agricultural planners. Where the efforts of the small farmer and the national authorities are closely and effectively linked by a network of local organizations, rural development does occur more rapidly.[33]

The function of agricultural extension is above all that of linkage, fusing national technological initiatives into unique ecological systems. Anyone who has ever tried to link the arrival of seeds from ministry headquarters with the start of the rains knows that this function is a frustrating, complicated, and demanding one.

One of the most debated questions of agricultural administration is where to place the focus of responsibility for linkages. The two alternatives most seriously advanced are (1) central initiative through vertically integrated crop authorities and (2) decentralized initiative through horizontally integrated local decision centers. Both of these institutional arrangements exist side-by-side in many developing countries. In East Africa the historical record favors the crop authorities. The more successful of them have had privileged access to international capital and markets, their own research establishments, special credit arrangements for their growers, and relatively large numbers of extension workers devoted exclusively to their crops. The Kenya Tea Development Authority even built its own feeder roads to get the small grower's leaf to the factory. When world prices are high enough to support such an apparatus, the narrowly focused attention of specialist decision makers whose careers ride on their crop assures a very effective degree of linkage.[34] Most crops have margins too limited to support their own exclusive agricultural infrastructure, however. Once such authorities become dependent on other agencies for supporting services, their specialized, vertical integration often becomes a liability in their attempts to make horizontal linkages. Furthermore, a crop authority is committed to the promotion of its particular crop, whatever its comparative advantage in a particular area, and therefore it often may try to pressure the farmer into an innovation which is not to his best advantage.

The early place of crop authorities in the affection of international development specialists has now been taken by "integrated rural development" at the local level. Guy Hunter summarizes the views of a wide, international network of development operatives and observers:

> In order to achieve optimum fit between local agricultural programmes and local potential, considerable authority for local programming must be delegated to local levels. Further, since there must be a degree of integration between agricultural and other social programmes, the wider planning process must give an opportunity for such integration at [the] local level.[35]

In their review of Asian rural development, Uphoff and Esman concur: "The more successful cases had engaged much more extensively in decentralization of operating decisions as well as local level planning." The horizontally integrated local decision center lacks even the potential autonomy of the crop authority, however, for it must sustain vertical links to central sources of support. Thus for decentralization to promote rural development well, it usually must operate within a framework of national controls.[36]

The present coexistence of local decision centers with national crop authorities or their functional equivalent—the specialist divisions of a central ministry—is untidy on organization charts but good for rural development. The crop authority or ministerial division is needed to provide crop-specific expertise and to negotiate the delivery of the requisite national contributions, such as finance, markets, credit, research, and favorable prices for inputs and produce. The local decision center is required in order to integrate the multitude of government initiatives in any area and to respond to location-specific opportunities and problems. These are different sorts of responsibilities. The problem is to assure that neither set is subordinated to the other and that their fulfillment is linked.

Our assessment of current Kenyan and Tanzanian practice suggests that local agricultural decision centers and national crop authorities or ministerial divisions should be independent of one another. Rather than one or the other being subordinate, they need to negotiate their relationships, under the overall control of national agricultural planners. In the ideal situation, each party has resources that the other knows it requires and therefore can bargain for an exchange of services. In a decentralized administrative structure, the national authority or division needs to sell to the districts programs to expand, improve, or maintain its crop. Conversely the local decision center wants to find the authorities or divisions that have the best agricultural packages for its particular structure of opportunities and problems. The interests of both are best served if they negotiate the amounts and timing of resources that each will provide, reaching a semicontractual agreement. Neither party would be responsible for the success or failure of a given project—both would be. Consequently they would have a mutual interest in renegotiating their agreements when the success of a program was at stake. Obviously such a bargaining approach would not always produce economically optimal decisions, sometimes because of incompetence, other times because one agency was narrowly pursuing its own interests at the expense of the nation's. The overall control of the negotiating process by national planners can eliminate the worst instances of suboptimizing, but ultimately we must recognize that we are working in a world in which planning mistakes are always being made. It is only that the mistakes of technically competent national decision makers derive from ignorance of local conditions and are therefore more difficult for the outside observer to identify.

The negotiating model of agricultural programing that we have sketched describes the underlying tendency in Kenya and Tanzania. Nonetheless both countries deviate from the model in ways that undercut their obtaining its full benefits. In Kenya central hierarchical controls over district agricultural decision centers are still too tight to give senior field staff adequate operational flexibility. The key problems relate to financial management. The views of local officers about development priorities for their areas are not seriously considered in making program allocations. Thus the district or divisional agricultural officer deals with crop authorities or ministerial divisions from a position of weakness; he needs their resources but can offer no assistance in obtaining increased finances. The senior field staff are also deprived of flexibility because the expenditure of funds is strictly controlled from the capital, even after allocation. Tanzania's decentralization has dealt effectively with both of these problems. A significant portion of Tanzania's development budget is administered through district and regional authorities. These bodies submit their own budget requests to the treasury and therefore have a small but independent influence on spending priorities. Once allocated, funds are dispersed to regional and district headquarters, where the entire financial control function is performed. The quality of accounting probably has declined as a result, but this loss has been more than compensated by the ability of local officers to use project funds when and where they are needed. The result has been an improvement in the speed and success with which local projects are completed.[37]

If Kenya's problem is too much hierarchical control of local initiative, Tanzania's appears to be too little integration of national support facilities into area-based programs. Once they have obtained funds, field officers in Tanzania forge ahead with implementation on their own and do not seem to obtain from the center sufficient expert advice, training facilities, diagnostic services, and the like. These problems of poor linkage also exist in Kenyan agriculture in communications and feedback. District-based officers need substantial support in these areas from national specialist agencies, as we pointed out in chapters 7 and 8. In a decentralized administrative structure the center needs to be every bit as strong as in a centralized one, but the orientation required is one of technical service rather than of hierarchical control. Tanzania and (to a lesser extent) Kenya have underestimated the importance of the local support function of a national headquarters.

Negotiations between national and local authorities on crop projects are particularly complex because the time of arrival as well as the quantities of inputs is so critical. A seed shipment, a staff training program, or a release of credit funds that misses the start of the rains might as well come next year as a week late. Early experience with Kenya's Special Rural Development Programme indicated that if damaging delays were to be avoided for projects with considerable interdependencies, the contributions of every field officer and central functional division concerned had to be carefully programmed. To accomplish this, Deryke Belshaw and Robert Chambers developed a set of procedures they called the Programming and Implementation Management (PIM) system. As spelled out by Chambers in his *Managing Rural Development*, the PIM process begins with a meeting of all the local officers concerned with a project, joined by representatives of the central agencies involved. The contributions that each party will have to make to the project and the time by which they will have to make them are discussed, plotted on a programing chart and agreed by those present. The time and quantity targets are set collegially but once agreed they become binding. The program of disaggregated targets then becomes the basis for official reporting on implementation progress throughout the fiscal year.[38]

The PIM system applies to implementation the kind of negotiating model that we favor for rural development decision making as a whole. Independent evaluations of PIM in the field indicate that it improves the quality of implementation programming and raises the commitment of field officers to their project work. The greatest problems with PIM derive from failures to apply it seriously. Some central agencies have sent only junior headquarters officers to negotiate their commitments. These staff lack the authority to bind their agencies, with consequent damaging failures in the delivery of inputs. If serious negotiation is to take place the delegates have to be able to make authoritative commitments, even if the bargaining process then takes two meetings, with an intervening recess for home office consultations.

The other disappointment with PIM has been that its effect on senior staff performance has declined as officers—both in headquarters and in the field—have perceived that its implementation reports are not being used in making promotion evaluations.[39] One of the potential advantages of an objective system of recording performance is its use in an incentive system. As we remarked in chapter 6, the motivation of Kenyan civil servants is undercut by the failure to use performance as the predominant criterion for promotion. An incentive system is as much needed for senior as for junior staff. The most forceful incentives grow out of promotions based on objective evaluations, such as those made possible by PIM and other Management by Objectives systems. Since these procedural methods are built on staff participation in the setting of performance objectives, they can be used to apply organizational pressure while still avoiding two of the usual problems of bureaucracy, inflexibility and authoritarianism. We must stress, however, that performance evaluation needs to be applied to central as well as local officers. A system that records failures to meet implementation targets can be used by the president's office to assess the performance of national agencies in their support functions as well as to evaluate local staff. The motivational effect is needed at both levels in East Africa.

A Review of Recommendations

Let us pause to draw together the many threads of recommendations we have woven. There should be two primary foci of agricultural decision making: the district team and the national crop unit. At the district level, agricultural planning and programming should come from the senior ministry staff operating as a group. As suggested in chapter 8, this collegial unit would organize its own diagnoses of the local small-farm systems and experiment with crop opportunities, problem solutions, and technical adaptations. Itinerant national experts would provide technical support to the local group in the performance of these functions. Out of the group's experience would come its priorities for agricultural development in the area. These priorities would be negotiated with elected representatives and other civil servants at the district and divisional levels and with national crop units. The crop authorities or specialized divisions of the Ministry of Agriculture would try to sell their crop programs to districts and divisions and in return would negotiate finances, credit, inputs, markets, and prices and would provide technical support in the form of expert advice, research, staff training programs, and so forth. Both the national crop units and the District Agricultural Officers would be largely autonomous but subject to the common oversight and discipline of the Ministry of Agriculture or national planners. The relationships between these two decision foci would be negotiated and the timing of their respective contributions jointly programmed through the use of a procedural system such as PIM. The implementation targets thus mutually agreed would become an objective and major factor in senior staff promotional evaluations both at the local and headquarters levels.

Field officers would organize the work of their subordinates by creating a network of farmer groups and committees, the responsibility for which would be divided among their junior staff. The supervision of extension agents would be delegated to Location Agricultural Assistants, loosely but systematically overseen by the divisional Assistant Agricultural Officer. The evaluation of junior staff would be based on fairly objective assessments of their handling of farmer groups as judged at meetings and demonstrations. To provide stronger incentives for agents, these evaluations would be the primary basis for decisions on promotions in a career system that offered meaningful opportunities for upward mobility, as suggested in chapter 6. The resistance of junior staff work groups would be defused by setting work targets in consultation with them and by nonauthoritarian, helpful leadership styles. Objective evaluation and consultative supervision should stimulate feedback from extension agents about the problems they are experiencing.

These recommendations could not be put into operation as they now stand. Many more details would have to be worked out; the objections of critical decision makers would have to be met with modifications as the recommendations were made genuinely indigenous; and limited trials of the reforms should be conducted where possible to work out the inevitable unanticipated problems. Proposals like ours can really be only suggestive, never definitive.

Nonetheless there are certain general principles that have guided our recommendations, the majority of which we would hope to find reflected in any reformed system of agricultural administration.

1. The details of an extension service's organizational structure have a profound impact on its performance and should be subject to continual, careful review and planning.

2. The effective linkage of national and local resources is critical to agricultural development.

3. All agricultural programs, even locally initiated ones, need expert support from national specialist agencies if they are to achieve their full potential. The types of needed support range from technological advice and marketing assistance from crop authorities, on one hand, to assistance with staff training or with agroeconomic analyses provided by functional divisions of the Ministry of Agriculture, on the other.

4. The communication of information across several hierarchical levels is a complex, demanding task and is best performed by specialists.

5. Senior field staff must have significant autonomy in their agricultural planning and programming if area-specific problems and opportunities are to be properly treated.

6. The arrival of inputs and the timing of the contributions of all those concerned with a project—both locally and nationally—should be carefully programmed.

7. The experience of small farmers should be continuously sought and diagnosed in order to identify area-specific opportunities and problems.

8. Recommendations and insights must be subjected to small-scale experimentation to test their appropriateness under actual local conditions.

9. A system of district and national seminars in which farm system diagnoses, small field experiments, and technological innovations are discussed collegially would promote a more competent and problem-oriented professionalism among senior staff.

10. Systematic provision for dealing with poor farmers must be made if they are to receive anything approaching their fair share of extension services.

11. Group methods are the most effective form of extension. In

addition, they facilitate service to poorer farmers and are readily supervisable.

12. Agricultural development programs are crisis-prone and constantly encounter unanticipated problems. Responsibilities therefore have to be based on results to be achieved rather than tasks to be done if the flexibility needed to meet the unexpected is to be maintained.

13. Nonetheless, when staff are judged by the attainment of objectives over which they have only marginal control, little if any sense of responsibility or incentive to perform is created.

14. Therefore, the results by which performance is to be judged must be those which a staff member has the capacity to produce. These results may be almost as broad as a program's objectives for those at the top of the organization but will be ever more narrowly defined, partial contributions toward the goals as one moves down the hierarchy.

15. Staff at all levels respond best to nonauthoritarian, helpful styles of supervision. As far as possible, staff should be consulted on their work targets.

16. Definable work targets make possible the objective evaluation of job performance. They therefore have a greater incentive effect and help to relieve tensions between supervisors and subordinates. When combined with helpful supervision, objective evaluation also stimulates feedback about work-related problems.

17. Promotions should be based largely on performance. Training, education, and capacity should be used only to define prerequisites for promotions and should not be a barrier for most employees.

18. The promotion opportunities for staff of all levels need to be plentiful enough to provide meaningful incentives.

19. It is often necessary to promulgate the exact procedures to be followed in executing an administrative function if the desired result is to be achieved. The negotiation of implementation schedules, the setting of work targets, the evaluation of performance, and the granting of promotions are examples of particular areas in which attention to procedural detail is beneficial.

Development Bureaucracy

There is an extensive literature with a well-defined position on the organizational features necessary for development administration and the majority of the principles outlined above are consistent with the tenets of this school of thought. Several of our recommendations, however, run contrary to its accepted wisdom. The conflict concerns our desire to tie rewards to definable performance criteria and to formalize, routinize, and depersonalize many decisions that are personalized and flexible at present. Our break with the development administration consensus on these points does not grow out of our own work alone. A number of those who have been thinking about the problems of administration in Kenya have come more or less independently to the conclusion that the situation requires more organizational planning and attention to the development of procedures. The experience of the Kenya Tea Development Authority with such a designed organization has been widely hailed as a success.[40] Robert Chambers has written about the "primacy of procedures" and together with Deryke Belshaw has conducted some generally successful field experiments with procedural innovations, some of which we discussed earlier in the chapter.[41] A high-powered Kenya government commission of inquiry came out strongly in 1971 in support of the use of Management by Objectives, one of several possible systems of detailed organizational planning and control.[42]

The contrary position of the development administration school follows analyses such as those of Tom Burns and G. M. Stalker, which suggest that there are two ideal types of modern organizational structure: the mechanistic and the organic. The mechanistic type of structure is seen as appropriate for routine conditions and is characterized by a stable specialization of tasks, a precise definition of role obligations and techniques, and a tendency for most operations to be governed by a hierarchical structure of authority and communications. The organic structure is held to be best for changing conditions and is characterized by a continual reallocation of tasks, a flexible definition of role responsibilities and methods, and a propensity for most work behavior to be directed by a purely collegial network of advice and communication.[43] Economic development obviously is concerned with change, and the environment in which government development agencies are working is imperfectly understood, requiring continual adaptations in program content and methods. Consequently, many scholars have concluded that the best structure for change agencies is the organic type.[44] So firm is this conclusion that it has become common to use the term bureaucracy for mechanistic structures and development administration for organic ones.[45] Any proposal for detailed structural planning and the use of organizational procedures in a development agency is therefore likely to unleash a torrent of challenges: "What is to prevent today's liberating procedures from becoming tomorrow's bureaucratic chains?'' ''Will not the rules become ends in themselves and inhibit the achievement of the basic goals?'' ''May not these procedures limit the administrator's perspective precisely when it needs to be widened to the full complexity of his environment?''[46]

There are several levels of response to these antibureaucratic challenges. First of all, it is false logic to argue that because development agencies are designed to create change, they are working in changing conditions. A ministry charged with road building can bring about major changes in a nation's infrastructure, but the work it does is highly routine. Similarly, it does not follow that because an agency needs to be adaptable in some of its tasks, none of its work is routine. Virtually all development agencies have categories of staff with quite stable responsibilities. Routinization of the management of these personnel (that is, making that part of the organizational structure mechanistic) can free the attention of decision makers for problems that do require adaptation. It is possible for some staff in an organization to be governed mechanistically while others' work is structured organically.

Hospitals, for example, tend to use different types of control for their nonprofessional and their professional staff. In like manner, different tasks performed by one individual can be subject to different management structures. For example, Herbert Kaufman makes clear in his study, *The Forest Ranger*, that a general framework of procedures and hierarchical controls does not hamper the ranger in his highly adaptive responses to an unpredictable forest fire. In fact, the United States Forest Service's system of controls works to guarantee that the individual ranger will have a variety of adaptive options open to him when a fire does occur.[47] The elaborate routine of military organizations is similarly designed to provide maximum flexibility of response in the crisis of battle.

Second, the advocates of development administration assume that a more mechanistic structure will be a more authoritarian, narrow, and rigid one. In the highly competitive environment of the Scottish businesses studied by Burns and Stalker, it may well be true that the only viable alternative to a mechanistic structure is a more organic one. In other environments, however, organizations may have settled into informal patterns even more authoritarian, narrow, and rigid than a planned and formally established mechanistic structure would be. For example, situations have been noted in France and the United States where formal procedures lead to a lessening of authoritarianism rather than to an increase.[48] Similarly, systematic accountability may entail dangers of rigidity or red tape in the provision of services, but it also reduces the likelihood of arbitrary or incompetent disregard for the interests of clients.

The dangers of procedures, careful definitions of tasks, and other ways of formalizing organizational operations are well documented.[49] Yet these devices have benefits in the circumstances of some development agencies that may outweigh or even be separable from their disadvantages. Procedures, definitions of responsibility, and such, may be thought of as having four basic characteristics: they are formal, sanctioned, systematic, and relatively permanent. All organizations have at least informal operating methods in order to achieve some predictability in participant behavior. One purpose of making procedures and definitions formal is to tie them directly or indirectly to the organization's sanctioning structure. Utilitarian organizations use sanctions to see that their employees pursue the organization's objectives. Rewards and punishments are the ultimate ways employed to maintain control over the behavior of subordinates and gain their acceptance of a task orientation. In many organizational contexts, employees are well aware that their long-range success or failure in their careers depends on their performance of their tasks. When a task orientation is firmly internalized in this way, attempts to detail an employee's responsibilities, to add to the sanctions governing his behavior, or to remind him of their existence may actually have negative effects by making him narrow or resentful in his work, hence the stress on an informal, nonauthoritarian administration in these circumstances. Operating procedures are usually only recommended in these situations and are not tied to sanctions by being formalized. Nonetheless there are other organizational contexts, and many of these exist in Kenyan development agencies, where employees either have ceased to believe that their careers depend on their task performance or feel that their promotion prospects are based on only a small portion of their total task.[50] In these circumstances, a reaffirmation of the sanctions governing the civil servant's behavior and a promise to apply them to the wider definition of his task are likely to increase and broaden his task orientation.

System in the division and execution of work, another characteristic of procedures and job definitions, helps to ensure thoroughness in the pursuit of tasks and to relate partial duties to one another. After a system of work becomes formalized and relatively permanent, however, changes or unanticipated problems may mean that the job becomes incomplete and the parts unrelated to one another. In such situations occur the well-known problems of inflexibility and of making ends out of means. That is the danger inherent in procedures. But lack of thoroughness and poorly coordinated work are also characteristics of *dis*organized work situations and may well be worse in these circumstances. Thus, the arguments for innovative, flexible administration that have been developed in highly bureaucratized societies are misleading when applied to underbureaucratized societies such as Kenya's.[51]

System also can be introduced in a decentralized and nonauthoritarian way, characteristics presently lacking in Kenyan administration. Earlier we mentioned the Belshaw and Chambers PIM procedures for organizing, coordinating, and regulating plan implementation activities on rural development projects. These procedures actually promoted collegial responsibility and field staff participation in decision making. What is more, by routinizing the reporting of delays and those responsible for them, the PIM system made it possible for subordinates to point out shortcomings in the performance of those above them and so provided a new device for field officers to pressure their headquarters over damaging holdups in authorizations.[52] If they are carefully designed to that end, procedures can achieve nonbureaucratic results, such as reversals in the hierarchical flow of demands and increase in subordinate participation in decision making.

Finally procedures and task definitions tend to take on an air of permanence. Once formally established, operating methods are not easily changed. They may become inflexible, making the organization unable to respond to changing circumstances. It is true that informally created operating methods also have a tendency to become permanent, but more damage is likely to be done by outdated methods in a formally established mechanistic system. Although some specific procedures, such as the Kenyan ones we have just discussed, have flexibility built into them, the level of an organization's total procedural system the problem cannot be bypassed. Nonetheless with effort it can be overcome. The first step is to appreciate that the details of any organizational structure need to be revised constantly if they are going to remain pertinent to the agency's operating problems. Kaufman indicates that the procedurally oriented United States Forest Service has recognized this need and has maintained the vigilance and openness to suggestions that are necessary to meet it.[53]

Perhaps one reason that developing countries are particularly susceptible to the rigor mortis of permanence is that they have tended to rely on foreign consultants in restructuring their government agencies. Since local personnel then are only marginally involved in the reform exercise, they are unable to make intelligent adaptations in the new set of procedures at a later point. As Jon Moris has remarked, the danger of organizations constructed by expatriates is that their features either will be perpetuated ritualistically or altered with uncomprehending destructiveness once the designers have packed their bags and left.[54] Structural reform of any kind is an ongoing exercise and governments such as Kenya's need to acquire a permanent and indigenous capability for it.[55] Unlike the other dangers of procedures, undue permanence is neither easily sidestepped nor acceptable in some circumstances. It has to be met and overcome through conscious effort. If the commitment to do so is accepted, however, we believe that procedural reforms promise great payoffs for civil services in a stage of organizational development similar to that of Kenya.

The basic point is that bureaucratization is not always a bad thing, even in the development field. There are routine tasks associated with development and these can be mechanistically structured to the great benefit of organizational goal achievement. There are other areas where the work is not stable but where, in Kenya, organizational planning and the use of procedures would be profitable. We shy away from wide reliance upon organic structures in Kenya, not because Burns and Stalker are wrong in their analysis, but because they assume certain conditions that often do not apply in Kenya. Their analysis assumes that employees have a broad commitment to the organization's goals. Burns and Stalker therefore expect that if an organic structure were to remove specific definitions of responsibility, the worker would accept an obligation to the organization's general goals and find a new and appropriate role for himself in their fulfillment. Burns and Stalker did their study on managerial-level staff in Scottish industries, where these assumptions seem reasonable. It is not at all clear that the same would be true of the factory hands in those industries. Such an internalized and generalized goal orientation certainly is not common among the field staff in Kenyan development agencies.[56] A mechanistic structure does not depend on such assumptions about employee goal orientation—hence our interest in it.

If mechanistic structures are to be termed *bureaucracies*, countries such as Kenya often have need of *development bureaucracies*. We use this hybrid term because our recommendations involve both organic and mechanistic elements. Aspects of the organic structure are needed—flexibility, programming by negotiation rather than hierarchical fiat, consultation in the setting of work targets, and an independent, collegial role for field officers in diagnosis and experimentation. We also see formalization as a necessary but dangerous means, to be carefully watched, and not as an end in itself. We do not share the belief of Chambers that procedural reform is the primary need in the Kenyan civil service, but we do accept that procedures must exist and require attention. The problem is that Kenya has achieved the efficiencies of neither the mechanistic nor organic ideal types. We contend that an intelligent system of organization that is formalized and relatively inflexible is preferable to a flexible disorganization that leaves work undone and uncoordinated. We believe that an intermediate stage in the development of the public service should be a period of relatively mechanistic *and intelligent* bureaucracy, which instills a sense of task orientation, a grasp of what are one's most important functions, and a feel for the amount of cooperation needed with one's colleagues. After these lessons of development bureaucracy have become a part of the general culture of the civil service, the highly flexible, organic management of the development administration school would be appropriate.

11 Organization Theory
in Kenya

A Review of Conclusions

This study has encompassed two separate but continually interacting objectives. The first has been to understand the particular organizational dynamics of the agricultural extension services in Western Kenya. In this way we have hoped to learn the basic social processes by which such services operate in developing countries and broadly how they could be manipulated to improve their contribution to economic development. Following these lessons we have made a number of organizational recommendations, which were reviewed in the preceding chapter.

A secondary objective has been to examine the extent to which the theories developed in the industrialized West for understanding formal organizations are useful in the context of rural, developing Kenya. Here we have wished to see how widely one part of Western scholarly enterprise could be applied. Now we need to examine our conclusions for their overall theoretical implications.

Individual Motivation

The basic insights that we have gained into the management of field agricultural extension agents in Kenya can be grouped under four headings: individual motivation, work group activities, responses to different styles of authority, and the impact of organizational structures. Concerning the first, we found levels of expectation to have a fundamental effect on the motivation of extension workers. Junior staff bring to the Ministry of Agriculture certain views about how employees such as themselves deserve to be treated. Objectively, all ministry field agents receive better terms of service than they would be likely to find on the open employment market. Subjectively, whereas those agents with only primary schooling have their expectations overfulfilled, those with secondary education find their expectations disappointed. Two of the most important reasons for this anomaly are: (1) the secondary-school graduate AAs are those most obviously disadvantaged by the colonially established salary scales, based originally on racial discrimination; and (2) those with post-primary education were the ones who benefited dramatically at Independence with the rush of Africanization promotions. The rapidity of advancement during the first years of Kenyan independence created levels of expectation for those with secondary education that could not possibly be sustained. The discriminatory salary scales are alterable but nothing except the cold water of bitter disappointment can lower the temperature of expectations to a level that can be met. Because at present extension agents with only primary schooling are highly satisfied with the ministry as an employer and those with secondary education are not, one finds higher motivation and better job performance among primary-schooled agents and low motivation and poor job performance among those with secondary qualifications.

It is clear that in the present circumstances Kenyan extension agents with secondary education cannot be satisfied with any conditions of employment that are feasible. The alternatives are then either to rely on primary-schooled agents or to develop a set of incentives that can overcome the inertia created by the dissatisfaction of those of secondary level. We believe that the existence of an attractive number of major promotion opportunities, stretched over the better part of an agent's career, can in fact stimulate sufficiently strong motivational forces to make the more educated agent the better performer.

The magnitude of the impact on job performance of disappointed expectations and inadequate incentives needs to be underlined. In every one of the four aspects of job performance that we examined the quality of work of those with secondary education was less than that of those with only upper primary schooling. This drop in the level of performance occurred even in those aspects of agricultural extension work for which more education should have been a decided advantage. The acknowledged contribution of education to increased competence in many fields cannot be considered without also taking into account its even more powerful and damaging effect on motivation in the present Kenyan circumstances.

Work Group Activities

The second set of insights concerns the significance of the peer group of extension agents working together in a location. We found that junior staff in the ministry have a sense of common identity that sets them apart from their senior staff supervisors. The interest-group conscious-ness of the ministry's subordinate staff finds organizational expression in the informal bonds uniting the agents working together in each administrative location. These work groups are not very cohesive, for their members come together only a few times each month. But some of them are strong enough to have a significant effect upon the way that the junior staff who belong to them do their work. Those location groups bound together with a network of friendship ties have members who are more strongly committed to a ministry career and are less likely to be dissatisfied with their jobs. The network of informal communica-tions in the group is also used to keep its members informed of the ministry's technical recommendations; those groups in which interac-tion between agents is easiest and most frequent are the ones in which Agricultural Informedness is highest. On the other hand, the existence of work-group solidarity gives the agents a collective autonomy that shields them from the demands of their superiors. Those groups with the best internal organization are able to protect their members from hierarchical demands and even to offer collective resistance when necessary. As a result, the groups that are strongest because of their small size and internal friendship cohesion do the least amount of extension work.

The informal organizations of extension agents have major effects on work productivity, but their existence is virtually unknown to superiors and outsiders. The sanctions that the senior staff are prepared to use against any uncooperative subordinates are so great that work group organization is kept virtually invisible. The strategy of junior staff is to organize, to protect one another, and to resist in strict secrecy, never even admitting what they are doing. To organize collectively is so dangerous that apparently it is done on a collegial basis, without any stable leadership—formal or informal. This whole pattern of work-group identity and activity is what we have termed anonymous group conflict. Such anonymity poses particular problems to the ministry, for it is very difficult to persuade groups that are clandestine and without visible leaders to use their collective strength for goal achievement rather than resistance to the hierarchy.

Responses to Different Styles of Authority

The form of supervision traditionally used in the Ministry of Agriculture is highly authoritarian, remote, and subjective. Our study has shown that it is also counterproductive. Those work groups that are under the most authoritarian and remote LAAs have the lowest levels of Visit Effort.[1] Such strict LAAs tend to spur their work groups into resistance against them. On the other hand, the agents in the locations with helpful and popular LAAs make the greatest Visit Effort. The social obligations created by these LAAs limit the desire of their subordinates to restrict their work. Similarly the AAOs who belong to the same tribe as their junior staff are less remote, inspire more confidence, seem more approachable, and consequently get more Visit Effort from their agents. Those AAOs to whom subordinates are able to go freely for technical advice also generate more Innovativeness and feedback from their staff. Furthermore, we saw how the exalted status of the AAO, putting him beyond subordinate question or challenge, has made him a major source of misinformation for his staff on technical matters. Finally, the subjective nature of the AAOs' decisions on punishments and promotions has made these sanctions unreliable and thus ineffective as incentives for agents.

Our evidence suggests that the most effective form of authority in the ministry would arise from a combination of personal warmth from the supervisors and impersonality in the organization. An objective system of evaluating agent performance and tying promotions to the evaluations would provide an impersonal set of high incentives to do good work. At the same time it would reduce the pressure that supervisors now feel to be authoritarian. Thus they could be more helpful and friendly to their subordinates, creating new social obligations to work hard, providing more leadership, and stimulating greater feedback.

The Impact of Organizational Structures

Inadequate or unintelligent attention to structural problems in the Ministry of Agriculture is the root cause of much of its inefficiency. The types of structural issues that we have in mind are not matters of the organization chart—who will report to whom and which ministry or marketing board will be responsible for what program. Reorganizations in this area are greatly overrated and have little impact on the way in which an extension service actually does its work.[2] Instead we are concerned with the day-to-day structures of incentives, authority, communication, and planning. The organizational pattern inherited from the colonial period exhibits minimal central control over senior field officers, leaving them to act as independent hubs of activity at the center of spokes of junior staff personnel. These field officers have been able to operate in a highly personal manner with relatively little structure or support supplied from headquarters. This was probably never a very efficient method of organization but it was effective when colonial officers were largely independent of national resources and of collegial cooperation. Such autocratic autonomy has vanished, due to the increased involvement of government in economic development. Likewise, the labor of the junior staff is now so expensive that it should be used efficiently, and the small farmer has reached a stage of agricultural development that demands better quality service from the extension agents.

The failure to create new organizational structures to deal with the changed circumstances of extension work has multiple ramifications. Well-trained and expensive agents lack sufficient incentives to work at anything like their full capacity. The highly personal and nonsystematic supervisory system not only fails to keep junior staff at work but alienates them as well, lowering effort further and smothering feedback and innovation. Even the extension service's central function of disseminating technical information is frustrated within the ministry itself by an ad hoc communication structure, which not only fails to keep extension agents adequately informed but which even occasionally disseminates misinformation.

The ministry clearly needs increased and ongoing attention to organizational planning. Improvements in feedback are required and much greater collegiality in decision making and consultations would be desirable. However, we also have argued that a carefully thought-out increase in the procedures and formality governing the field administration of extension activities is needed. At this stage of development, a partially bureaucratic structure that was intelligently constructed and periodically reexamined would dramatically increase the productivity of Kenya's extension services.

The lessons that we have drawn out of our study obviously apply primarily to the Kenyan Ministry of Agriculture, but many other extension organizations and rural development agencies in poor countries have broadly similar problems. The insights drawn from this study would apply to them as well. Throughout our analysis we have tried to avoid being too specific, precisely in order to offer the most general applications possible.

A Review of the Utility of Western Theory

Throughout this study we have examined the particular characteristics of Kenyan extension administration in broad theoretical perspective, not only to derive the most general possible practical lessons but also in order to discover the types of theories that work well for other Kenyan organizations. We have been anxious to assess the utility of Western organization theory in an African setting in order to gauge how far it can be used as a guide to organizational policy throughout the developing world.

The amount of Western organization theory that we have been able to examine is necessarily limited. The term *organization theory* is used to cover both the work behavior of the smallest employee groups and the power struggles of whole organizations. Obviously no single study can cover so many different levels of analysis and foci of interest. This study has been defined by the point of convergence of three interests. First, we have been concerned with a utilitarian organization, one that uses material rewards in order to gain the cooperation of its members.[3] Second, we have concentrated on the behavior of subordinate staff, those on the bottom two steps of the hierarchical ladder. References to those in the middle and upper echelons and to the types of tasks that they perform have been largely peripheral and subservient to our central purpose. Third, we have focused on the aspects of the organization that explain variations in its productivity. We have attended to issues such as the distribution of power or organizational maintenance only as they bear on the problems of productivity.[4] Thus we have examined that part of organization theory that treats the effects of subordinate behavior on productivity in utilitarian organizations. As it happens, the area of our concentration coincides with what is probably the most fully developed and rigorous part of organization theory, having behind it as it does the vast amounts of industrial sociology and small groups research.

Western organization theory proves to be a remarkably useful tool for understanding the effects of subordinate staff behavior on productivity in Kenya's Ministry of Agriculture. Throughout the preceding chapters, we have discussed the extent to which particular findings conform to the hypotheses offered in the literature. In only a few instances did we find differences. We were able to confirm the usefulness for Kenya of existing theories about job motivation and about worker responses to styles of supervision, for example. At one stage in our work, we culled the Blau and Scott review of the organization theory literature for hypotheses about groups that could be directly tested with our data. Of the thirty-six we found, twenty-one were supported by our findings and in only five cases could we say with any confidence that the relationship hypothesized by Blau and Scott probably does not exist for these workers. In no case could a relationship opposite to a Blau and Scott hypothesis be established with statistical significance.[5]

Furthermore, without the Western literature we probably never would have discovered one of the most important characteristics of extension a-gent behavior in Kenya. Like most other outsiders—both African and European—we were deceived by the apparent subservience of the junior staff and did not realize that they were engaged in collective restriction of their output. Our first clues to the reality of anonymous group conflict came from patterns in our data that the literature identifies with collective influence on output. For example, Stanley Seashore has found that when groups are setting production standards, there is more variability in output between the most cohesive groups than there is between the less cohesive ones.[6] We discovered that the more cohesive agricultural extension groups also showed more variation in their average Visit Effort than did the uncohesive ones and thus immediately began to suspect that the work groups were setting output standards.

Nonetheless, the great usefulness of the Western hypotheses to us and their standing up well in our analyses should not blind us to the instances where they did not work. We have already mentioned that a seventh of the relationships that were hypothesized by Blau and Scott and that we could test with our data apparently do not apply in the Ministry of Agriculture. In chapter 10 we also found inapplicable to present-day Kenya the proposition that organic organizational struc-tures are most appropriate to development agencies. Finally, we were totally unsuccessful in applying Theodore Caplow's SIVA model of organizations to our work group data. This failure is all the more notable because Caplow himself claims universal, transcultural validity for his model.[7] Our data simply could not be made to fit into a rigorous, quantitative statement of his predictions in any meaningful way.[8]

It would have been naive to expect any investigation such as ours to lead either to a wholesale rejection of Western theory or to its simple acceptance. Certain pieces will be found helpful, others misleading or useless. Western theory has not been and will not be found to be so universal that it can be applied to organizations in developing countries without further research. But we are now convinced that it is grounded in sufficiently basic principles of human behavior that its use as a guide to research will lead more quickly to a valid body of organization theory for any particular region of the world than would a completely fresh start. We need now to analyze the experiences with Western theory that we have had in this study. We must identify the underlying reasons that have led us to accept or reject certain theories and discuss the degree to which organization theory is likely to fit other Kenyan organizations as it has the extension services.

Contingent Relationships and Culture

There are a number of Africanists who reject the use of any Western theory, arguing that what is most important to behavior in Africa is that which is peculiar to the continent.[9] These arguments often move from the demonstrable fact that Africans frequently do not act the same as Europeans to the assumption that the basic mechanisms governing their responses to the world are different. The case is a plausible one, but it seems to us that an alternative and equally reasonable assumption would be that the distinctive features of African behavior derive from an essentially universal mode of response to the peculiarities of the African environment. There was a time when the non-Western peasants of the world were thought to be economically irrational, to respond to economic stimuli in a way totally incomprehensible to the Westerner. More recently, economic analysts have come to the conclusion that once proper account is taken of certain basic and distinctive features in the peasant's economic environment, theoretical models of his behavior can be constructed that are sufficiently transcultural to be applied to similar conditions in the West, Asia, and Africa. The problem has not been to construct a new, non-Western body of theory but to appreciate the importance of contingencies that are significant in developing countries but have been ignored in the West.[10]

The social sciences deal almost exclusively with contingent rather than sufficient relationships. This simply means that instead of two factors always being related to one another (a sufficient relationship), they are related to one another only under certain conditions (a contingent relationship).[11] Sometimes the factors upon which a relationship is contingent are indicated explicitly, sometimes they are left implicit, but there are virtually always some conditions upon which the existence of a relationship depends. In chapter 5, for example, we found that there is a tendency for worker productivity to decline under relatively authoritarian supervision and that this tendency appears to be cross-cultural. This does not mean that the reported relationship is a sufficient one but only that it does not seem to be contingent on the features of one particular national culture. We explicitly stated that one condition for the existence of the relationship is that the organization be a utilitarian one. The proposition might not apply if the organization did not have a hierarchically ordered set of authorities who controlled the disburse-ment of material benefits and if the workers were not motivated primarily by those rewards. Our statement of the proposition also was based on the assumption that the effects of other important influences on worker productivity were being held constant. The regression equation used to establish the probable validity of the proposition included the interaction effect of group size and friendship cohesion. In effect, the multiple regression procedure enabled us to create a hypo-thetical situation in which the Size-Cohesion Interaction Effects for all the groups would be equal and then to see the remaining effect of supervisory style.

Our example of a contingent relationship illustrates two types of qualifying conditions, the difference between which is important for a discussion of the cross-cultural application of theories. The first is the all-other-things-being-equal type of condition, which appears in multi-ple regression equations and similar statistical exercises. We will call these *control conditions*. Logically this type of contingency would be expressed: If X is held constant at any value level, then a change in the value of Y is associated with a change in the value of Z. The same relationship between Y and Z exists in all conditions but it will be clearly visible only when the effect of the stated condition (X) is held constant. The second type of contingency is the only-in-certain-situations sort, which we will call *preconditions*. Logically expressed, this would be: If and only if X has the particular value b, then a change in the value of Y is associated with a change in the value of Z. Here a quite different relationship between Y and Z may be found when the condition X is something other than b. While a control condition would be handled in multiple regression by simply adding the condition as another term to the regression equation, a precondition would require the construction of different equations for each value of the precondi-tion. Consequently, we tend to regard discoveries of control conditions as simply adding refinement and complexity to existing theories, where uncovering preconditions is seen as limiting existing theories and requiring the development of new theories to fit the alternative values of the preconditions. Propositions of relationship that are hedged only with control conditions could be termed universal while those with preconditions are situationally bound.

It is possible for the sociological theory describing a relationship to have cross-cultural validity although the relationship is powerfully conditioned by aspects of national social structure and cultural values. A good example of a proposition that has cross-cultural validity but is nonetheless conditioned by societal values is the one we have just been discussing from chapter 5. We found that worker productivity tends to decline under supervision that is more authoritarian than the workers in that particular culture and type of work have come to expect. Cultural values condition workers' expectations and these in turn determine the point at which more authoritarian supervision will lead to a decline in productivity. The relationship is therefore contingent on culturally derived expectations, but the contingency is of the all-other-things-being-equal type—it is a control condition. Thus the theory has cross-cultural applicability as long as the condition of expectations is placed upon it.

The contingencies to which we must be particularly sensitive in the cross-cultural use of theory are preconditions, that is, those of the only-in-certain-situations type. If a relationship is dependent on a precondition that is virtually always met in the society in which the theory was developed, it is quite possible that the contingency has never been noted. In this case the theoretical literature would contain no warning that the theory is inapplicable in societies where this particular precondition is unfulfilled. In order to anticipate difficulties in such a situation, one needs to analyze the theories for their likely hidden preconditions and then to consider whether and when those preconditions are met in the other culture. This approach involves taking an analytic approach to the effects of culture and society on sociological relationships. Just as a cook singles out altitude as the salient part of the environment when writing a cake recipe, an analyst needs to identify the individual features of a social setting that are significant for a particular social theory.

Authority Cleavage as a Precondition to Organization Theory

The theoretical analysis of this study began in chapter 3 with a consideration of conflict groups based on the distribution of authority. One of the reasons for beginning there was that the propositions of Western organization theory may well be contingent on authority's being the dominant source of cleavage within utilitarian organizations. As it happens, conflict groups in Kenya's Ministry of Agriculture are indeed based on the division between those who have and those who do not have authority. Hence our study can cast no light on whether the predominance of such a cleavage is in fact a precondition for the applicability of much of organization theory. Nonetheless there are good reasons why such a contingency is likely. Authority recurs repeatedly in organization theory propositions—responses to supervisory authority, the effects of hierarchical status on the upward flow of information, the protective organization of those without authority, the greater credence given to information disseminators of higher status, and so forth. Implicit in many of these propositions is the idea that the distribution of authority is the dominant social fact in organizations. Were we to have a situation, however, in which one group of both workers and managers were united in conflict with another group which also cut across all authority lines, it seems likely that much of organization theory would be inapplicable or would explain only insignificant details of what was happening. Obviously it would be difficult if not impossible in such a situation for the workers to develop a protective organization against managerial demands. Similarly the flow of information would be influenced primarily by conflict-group membership, and information relevant to the conflict probably would pass up the hierarchy within each group fairly freely. Thus it is quite plausible that organization theories are contingent upon authority's being the dominant source of organizational cleavage. Additional research is needed to determine whether or not this is true.[12]

If it is true, what would this mean for the usefulness of its propositions in Kenya-like societies? In many developing countries the line between those with and those without authority is frequently and powerfully crosscut by another important cleavage—that of ethnicity (which in Africa now most often means tribal identity).[13] If organization theory were conditional upon authority's being the dominant source of intraorganizational conflict and if ethnic cleavages were actually dominant in the case at issue, the proposition would not apply. The usefulness of organization theory to societies in which ethnic conflict is important therefore may well be dependent on the relative strength of these two sources of conflict in organizational settings.

We have evidence from several studies on the relative importance of ethnic and authority cleavages in Kenyan organizations. Since this evidence bears directly on how widely applicable the findings of this study may be, we will summarize the studies here and attempt to draw a common set of conclusions from them. In the case of the Ministry of Agriculture in Western Province, we noted that the junior staff almost all belonged to one tribal group and that the majority of senior staff were outsiders. Thus the two sources of cleavage—authority and ethnic group—generally reinforced one another. In cases where the senior staff belonged to the same ethnic group as their subordinates, the cleavages based on authority were much less deep. This was particularly true of the junior staff, who seemed to lose much of their sense of distinctness when faced with supervisors belonging to the same tribe. They nonetheless still maintained informal conflict groups based on the work group. Authority or interest-group consciousness was much stronger among the superordinate staff, and among them a sense of separateness and superiority remained even when they belonged to the same tribe as their subordinates. This group identity among the senior staff was probably one of the pressures leading to the formation and maintenance of junior staff conflict groups.

Isaac Gobanga has written about a small bank in Western Kenya in which ethnic group seemed to be the sole determinant of conflict group formation. The bank's sixteen Luhya employees were divided into two neat cliques, both reaching from the managerial level right down to the messengers. The line of division between the two could be accounted for perfectly by lineage alliances. The formation of these two cliques seemed to have been initiated by the manager and accountant, who were engaged in a power struggle. The manager had gained his allies by promoting those in his ethnic group, while the accountant was promising protection and future promotions to those in his. Thus the suppression of authority as a basis of cleavage and its replacement by ethnic group was initiated by those in authority themselves.[14]

Dominic Muigai and Solomon Njooro have examined the Government Mechanical Workshop in Central Province, where the pattern of alliances was exactly opposite to that described by Gobanga. In this case the organization's forty-one employees were all members of the Kikuyu tribe, although from different districts. The cleavage between the four supervisory personnel and the thirty-five mechanics was deep and was not crosscut in anyway by district or other subtribal alliances among the staff. As with the Western Province bank, however, those in authority seemed to have created or at least to have sustained this particular pattern of cleavage. The supervisors went far out of their way to avoid any informal interaction with their subordinates and insisted on a strictly formal set of relationships.[15]

A further case in which authority determined the patterns of conflict is provided by Jeremiah Mukuha. He analyzed a large Nairobi bank in which two different racial groups were represented at both the supervisory and clerical staff level. It was clear that these racial differences were important to the staff in their social relationships, yet the tensions associated with the authority system in the bank were so great that the normal pattern of conflict and solidarity was along authority and not racial lines.[16]

Finally, we have an interesting sociometric study done by M. S. Ng'ang'a and B. M. F. Mwangola. In a section of the government bureaucracy in Nairobi, they found evidence of alliances across hierarchical lines based upon tribe. These alliances were weaker than had been expected, however, and the basic reason seemed to be that senior staff were not being as helpful to the subordinates of their tribes as the latter expected them to be. As a consequence it seems likely that authority was the dominant source of cleavage for the majority of subordinates.[17]

The above few cases seem to us to point toward a common conclusion, despite the several patterns of conflict manifested. Both authority and ethnic group are readily available as organizing principles for conflict. The actions of the superordinate group in the organization determine which of the two will predominate. If they choose to emphasize hierarchical differences or to exacerbate the normal tensions associated with authority, conflict groups will form along these lines. This was the pattern in the Nairobi bank and the Central Province mechanical workshop.

If those in authority wish to build their alliances along ethnic group lines, that too can be achieved readily, as was evidenced by the Western Province Bank. Race, tribe, and lineage are natural methods of distinguishing "us" from "them" in Kenya. These concentric ethnic groups represent firmly established and legitimated systems of social exchange. The Kenyan worker expects to receive special assistance from members of his ethnic group and in return is expected to provide help to others in his group whenever he can. In a situation of choice, he would expect to help and be helped by the individual who stood closest to him in the concentric circles of family, lineage, subtribe, tribe, and race. On the other side, the worker expects no assistance from those outside his ethnic group. Yet if a person in authority wishes to construct an ethnic loyalty group, he must meet the expectations of his allies. This means that any benefits he controls will have to be allocated to the members of his ethnic group.

As the number of benefits provided by supervisor to members of his ethnic group declines, so will the support he derives from the subordinates in that group. The study done by Ng'ang'a and Mwangola in a section of the government bureaucracy illustrates how subordinates' tribally based expectations are easily disappointed and their support diverted to outsiders who do help them. When the superior provides no rewards on an ethnic group basis, he may retain some residual ethnic group loyalty, but not enough to prevent the formation of conflict groups on authority lines. This was the situation that we described in the Ministry of Agriculture.

The above observations about ethnic group and authority in business and government organizations can be strengthened by Richard Sand-brook's analysis of factionalism in the Kenyan trade unions. Sandbrook observes that factional alliances of members and leaders are usually built on contractual rather than moral or ideological ties:

> The most common moral ties are kinship, ethnicity and friendship, [but] a common tribal background is seldom a sufficient source of group cohesion, without the promise of personal rewards.... [Tribe] serves to distinguish among workers "us" from "them." But there is no automatic response on the part of branch officials or members to appeals to ethnic loyalties. Exceptional leadership can negate tribalism by ensuring that all positions and benefits, both within the union and, as far as possible, within the industry with which the union deals are shared equitably among all ethnic groups. Alternatively, a strong leader may allow the tribes within the union to find their own balance by remaining neutral in most individual contests for branch and national office.[18]

Workers in Kenya give their loyalties in return for tangible benefits. They can easily be mobilized along tribal lines if they are attracted or threatened by an ethnically unequal distribution of benefits. But the benefits they have received or lost (or believe they will in the future) must be material ones; symbolic rewards, such as being led by one of "us," will seldom be sufficient in themselves. Appeals to ethnic loyalties without the promise of tangible benefits will most often be rejected, especially when the alternative leadership is offering an equal distribution of rewards. In the trade unions, clear ethnic alliances are rare, for seldom does one tribe control a majority of the votes. Instead, the successful leaders carefully balance the ethnic distribution of their slates of candidates and work for equal allocation of benefits. Balance is the strategy that favors the interests of the leaders in this setting and so they work to avoid divisive sectionalism and ethnically based exchanges of benefits.[19]

To return to the utilitarian organizations that are our focus of interest, ethnic group can be seen as the perenially open option for conflict group formation, with authority divisions as the default position. Ethnic loyalties are almost always available for exploitation by an organization's leaders. But use of them demands that benefits actually be distributed predominately along ethnic lines. Very few superordinate staff either are able or willing to allocate rewards so narrowly. Consequently ethnic alliances are usually only one of the bases of conflict group formation in the organization. As the exploitation of ethnic loyalties by superordinates becomes weaker, authority increasingly takes over as the source of cleavage. Since authority is already distributed unequally in the organization, no special effort is required by anyone for it to become a source of conflict. Much more research on the subject is necessary, but the evidence we have collected so far suggests to us that inequality in authority is usually the dominant cause of conflict group formation within utilitarian organizations in Kenya today.

As Ralf Dahrendorf has argued in *Class and Class Conflict in Industrial Society*, the difference between those who have and those who do not have authority is sufficiently fundamental that it is as natural a principle for division and conflict as any other aspect of human organization. It seems likely that authority is usually the dominant source of cleavage within utilitarian organizations even in societies where other types of cleavage are prevalent. Thus organization theory probably has more cross-cultural utility than we might otherwise have imagined.

Other Preconditions

Another likely precondition for the application of organization theory is that the formal authorities largely control the income which the employee derives from his position. Max Weber regarded the organizationally provided salary as one of the defining characteristics of modern bureaucracy.[20] The organization's sanctioning powers are much greater when its staff members derive their basic income directly from the organization and not from private sources or from service fees that they collect themselves. All modern states now pay salaries to their civil servants from government treasuries, but in many of them bribes provide employees with an independent and significant additional source of income. It seems likely that the effects of the formal authority system on administrative behavior would be almost obliterated where bribery was important and beyond hierarchical regulation. (If implicit

authority were given to subordinates to take bribes in certain circumstances and if they could be removed from bribe-taking positions if they
defied hierarchical instructions, bribes would be a part of the material
benefits granted and controlled by the organization. In these circumstances bribes would strengthen rather than weaken the effective power
of the formal authorities.) As we noted in chapter 9, Ministry of
Agriculture employees do receive private payments for their services on
occasion and these are not hierarchically regulated. Nonetheless bribery
seems to be a relatively minor determinant of extension agent behavior
in Kenya. Pressures coming from the ministry's authority system explain
a much larger part of junior staff behavior than does extralegal income.
Thus we found that organization theory can be applied even when some
bribery of staff exists. There are a few government offices in Kenya,
however, in which unregulated bribery is probably a major force.
Organization theory would be unlikely to be helpful in explaining
behavior in these offices unless its propositions were significantly
modified. Nevertheless, in most of Kenya's rural development agencies,
extralegal exchanges have a relatively minor effect on employee behavior, the authority system is predominant, and so organization theory
would be applicable.

We did discover other preconditions, however, which do limit the
applicability of parts of organization theory to present-day Kenya. In
discussing the arguments usually advanced for favoring "development
administration" over "bureaucracy," we criticized the failure of some
writers to specify sufficiently the conditions under which their case
would apply. Sociologists such as Tom Burns and G. M. Stalker have
documented the advantages of unstructured "organic" organizational
patterns over structured "mechanistic" ones when one is dealing with
major development activities rather than with routine production
ones.[21] Some scholars have rather uncritically argued that since public
administration in poor countries is basically concerned with promoting
economic development, it should move away from bureaucratic or
mechanistic structures and towards development administration or
organic ones. Yet some aspects of public administration are basically
routine, even when they touch on economic development. In the Burns
and Stalker analysis, these aspects would be most efficiently administered within a mechanistic structure.

Another problem with the simple transfer of development administration theories is that their rationality has been demonstrated only in
highly structured and mechanistic organizations in which there is a very

high degree of task orientation. In Kenya today many organizations suffer basically from disorganization and weak job motivation at all hierarchical levels. We cannot see how the flexible disorganization of development administration can improve this situation. Instead we have argued in chapter 10 that an intelligent mechanistic or bureaucratic organization with consultative features is the more appropriate response to the problems facing Kenya at this particular stage. At a later point, organic or development administration may be able to supersede this largely mechanistic structure and produce a still more effective system. There is no question of the theory of Burns and Stalker being proved wrong in Kenya, of course: rather, the theory deals correctly with a set of problems that are not important in Kenya at this time but might well be significant later. The path of theoretical development is to specify more and more carefully the conditions of applicability as we have done here, thereby moving to greater precision and deeper levels of understanding.

In our review of the empirical tests that we have made of organization theories, we noted that Theodore Caplow's SIVA model was not useful in explaining our data. On reflection we think that this was due to the character of Caplow's theories rather than to its Western origin. Instead of specifying the conditions of applicability of organizational propositions more precisely, Caplow attempts to develop fundamental laws of behavior for all types of small groups. In so doing he moves to a level of abstraction that makes the model difficult to apply to a concrete situation and misleading in some ways when used on the problems of particular organizations in a specific setting. There is a good chance that Caplow's model would not have stood a test in an American industrial setting either. Our experience with Caplow's model lends weight to our conviction that in cross-cultural research, theoretical depth of understanding is better reached through the careful specification of the contingencies under which the propositions are expected to operate rather than through the development of highly abstract "laws" of behavior.[22]

The Flexibility of Exchange Theories

The bulk of the contingencies that we have identified in this study have been of the all-other-things-being-equal type, that is, control conditions. The most important of the culturally influenced control conditions which we have found have had to do with levels of expectation. The account that we took of worker expectations concerning supervisory authoritarianism and promotion prospects was central to our analyses. Since values such as expectations vary widely between cultures, only theories that can deal with value differences easily are likely to survive in cross-cultural analyses. Economics has been the most successful of all the social sciences in developing a body of theory that has needed only adaptation rather than fundamental reform when applied to non-Western societies. It seems to us that the widespread applicability of economic theory is grounded in two attributes: (1) Its core concept— economic exchange—is a process that occurs in all social settings; and (2) it has a built-in mechanism for dealing with variations in values between and within cultures. The worth of everything in society is gauged by the amount of money one is willing to forego or to spend in order to obtain it.[23] Thus the economist can simply assume that status, loyalty, and silk kimonos have a consumption value in Japanese society that they lack in the United States and proceed with his analysis of the processes of economic exchange. The social theories that have the greatest cross-cultural utility are likely to be the ones with properties similar to these of economic theory.

Several of the theories used successfully in this study are grounded in the process of social exchange and have a built-in relativity about societal valuation. The exchange principle is given different names by the various theorists whom we have used, but its basic similarity to the principle of economic exchange is evident in all of them. James March and Herbert Simon call it organizational equilibrium and define it: "Each participant will continue his participation in an organization only so long as the inducements offered him are as great or greater ... than the contributions he is asked to make."[24] Victor Vroom uses the term valences for the strength of a man's motivation to perform particular acts. He argues that these depend on (1) the balance of satisfactions and displeasures he anticipates deriving from this and alternative acts and (2) the strength of his experience that these outcomes will in fact follow from this act.[25] The second element of Vroom's conception is clearly analogous to the principle of investment risk in economics, just as the balance ideas of March, Simon, and Vroom are direct equivalents of market exchange. Peter Blau's conception of social exchange presses into levels of analysis not touched by economic theory while still working with the same basic type of process. "A person for whom another has done a service is expected to express his gratitude and return a service when the occasion arises."[26]

The second necessary condition for success fulfilled by the above theorists is a certain relativity about the values with which different rewards are invested. March and Simon speak of inducements to the worker being "measured in terms of *his* values and in terms of the alternatives open to him."[27] Peter Blau uses the crucial concept of relative expectations: "The satisfactions human beings experience in their social associations depend on the expectations they bring to them as well as on the actual benefits they receive in them."[28] The related ideas of relative deprivation and relative benefit were extremely important to us in this study in explaining the patterns of worker motivation and response to authority that we observed.

We believe that theories of social exchange coupled with the concept of relative expectations will prove powerful in analyzing organizational behavior in non-Western settings. Even when applied to problems that have no counterpart in the West, the theories are remarkably useful. For example, M. S. Ng'ang'a and B. M. F. Mwangola have used Peter Blau's exchange framework to explain why Kikuyus in a Nairobi office show more loyalty to superiors of their tribe than do Luos. Both groups expect to receive special favors from leaders of their own group, both actually obtain less help from their tribemates than they had expected, and both receive somewhat more assistance from outsiders than they had expected. The Luo subordinates respond to this situation by deserting the Luo superiors who disappoint them and giving their support to the non-Luo superiors who exceed their tribally based expectations. In the same situation, however, the Kikuyu subordinates remain more loyal to their Kikuyu superiors. Ng'ang'a and Mwangola see this different response as arising from the Kikuyu Mau Mau uprising. The experiences of seeing Kikuyus betray and kill one another in the conflict gave them both much lower expectations of their tribesmen and far greater appreciation of the value of tribal solidarity.[29] Theories like Blau's offer a set of tools for identifying the general processes underlying even the response of a particular set of actors to a specific history and environment.

Conclusions

In conclusion, we have been pleasantly surprised at the utility of much of organization theory for explaining administrative behavior in Kenya's Ministry of Agriculture. Contrary to our first expectations, authority usually does predominate as the major source of cleavage within Kenyan utilitarian organizations, even though quite different types of cleavage prevail in the larger society. Since inequalities in authority characterize modern utilitarian organizations wherever they are found, this natural source of conflict is transcultural and, to the extent that it predominates, tends to create many further commonalities in the ways that these organizations operate. Of course Michel Crozier's seminal work in France has demonstrated that there are differences between cultures in organizational behavior as well, and we have shown the same for Kenya. It served Crozier's purposes to accent the dissimilarities between societies and it has served ours to stress the commonalities. Each of us could have written our studies with the emphasis of the other. Crozier would agree with us, however, that despite the variations between nations, many aspects of organizational behavior can be accounted for by a common body of organization theory.[30] Since the core of utilitarian organizations is their reliance upon authority and material rewards, the theories that deal with exchanges of rewards and with power and authority are those that are likely to have a good deal of cross-cultural utility.

The amount of knowledge about formal organizations in nonindustrialized societies is still grossly inadequate. Thus organization theory inevitably has certain cultural biases and weaknesses, as has been illustrated occasionally in this study. Nonetheless, it seems to us that enough of organization theory has penetrated to transcultural levels of explanation that it deserves to be tested outside the West rather than to be simply disregarded as culturally bound. Not only would such a strategy be more efficient for the growth of knowledge but it also would enable us to recommend more rapidly practical steps for improved organizational management in the developing world.

12 Reflections on the Political Environment of Agricultural Extension

Can the State Promote Agricultural Development?

Does a low level of social and economic development dictate the existence of a public administration too weak to be an instrument for accelerating development? We posed this question in our opening pages and then rushed on to consider the problems of managing an organization of agricultural extension agents. It is fitting that we now turn back to reflect on that question in the light of our study.

We should begin by acknowledging that, although the Kenyan economy is poor, it is developing at a respectable rate.[1] Thus Kenya's Civil Service is not to be made a scapegoat for any lack of progress. Instead we want to ask to what extent the public services have been instrumental in achieving the growth that has occurred and to what degree they could be used to speed up the development process even more. Kenya's prefectorial Provincial Adminstration has been crucial, of course, in assuring that some of the preconditions of economic development are met. In the years immediately after the 1964 achievement of independence, the Provincial Administration played the central role in the control and channeling of the separatist and populist threats to the new state. It thus provided the framework of political stability within which the strong capitalist forces in the Kenyan economy could grow.[2] Despite the contribution that peace does make to development, however, such law and order functions are not what we or the civil servants themselves have in mind when they talk about "development administration."[3] The real question is not whether the Civil Service has provided an atmosphere in which development is possible but whether it has been a positive force in economic growth.

To put the question within the framework of our study, have Kenya's agricultural extension services added to economic growth in the post-Independence years? In some sectors the answer is very clearly yes. The most dramatic contributions to agricultural development have been in corn and tea. Between 1964 and 1968, Kenya's corn crop increased by 160 percent. This increase was due almost entirely to marked increases in average yields per acre, based on the use of improved hybrid and synthetic seeds developed by the government's agricultural research stations.[4] In the small-farm areas this corn innovation was disseminated almost exclusively through the extension services. During the same first four years after Independence, tea production increased by 50 percent, expanding almost entirely on small holdings.[5] Growth in this sector was especially impressive because previously all tea had been grown on plantations and earlier attempts at smallholder tea development in India, Sri Lanka, and Java had failed.[6]

On the other hand, the Kenya government was unable to meet its agricultural growth targets in other areas, most notably for coffee and cotton. The slight drop in coffee production was due to a major decline in the world price and to a serious outbreak of coffee berry disease.[7] It is to the credit of the extension services that this disease now seems to be under control. The failure of the cotton expansion campaign was more serious and resulted primarily from extending the crop in areas where it was only marginally profitable for the average smallholder.[8]

An analysis of the Kenya government's successes and failure in agricultural development reveals some important points. The first is the great dependence of agricultural development on the dramatic profitability of new crops and technological innovations. The African smallholder responds as vigorously as any farmer in the world to opportunities that he can see to be profitable. But since the African small farmer does not keep any kind of accounts, it is very difficult for him to notice the value added by innovations that produce a respectable but small additional profit (say, 15 percent net). Only very large profit margins are visible to a farmer whose calculations involve nothing but very rough guesswork. Hans Ruthenburg has estimated that profits in the vicinity of 100 percent are necessary for the rapid and easy adoption of innovations on small farms in East Africa (that is, the value of the additional crops produced must be double the cost of the additional inputs required by the innovation).[9] The need for such high profit margins (at least at the time of introducing an innovation) puts great constraints on rural development. Expansion of a crop is dependent on favorable world market conditions. When a crop's prices drop to a level that is only just profitable—as is the case with robusta coffee and cotton in Kenya—it is extremely difficult to get farmers to increase production of that crop, even though it is in their own and the nation's economic interests for them to do so. An effective extension organization would be able to persuade farmers to adopt practices that have substantially lower returns than Rutherburg's 100 percent. But when profit margins on an innovation are low (perhaps below 20 percent), even the best extension service will be unable to get more than a few small farmers to change their ways.

Agricultural Research and Farmer Demands

Unless a crop is in short supply or dependent on some unusual ecological conditions, the only way a country can maintain high profit margins for it is to employ one of the more efficient methods of growing it. Thus in the long run nations such as Kenya can expand their income from agriculture only through developing and using more efficient husbandry technologies. In today's world, a nation's ability to expand the agricultural sector of its economy is dependent on the quality of its agricultural research establishment.

One of the reasons for the strength of Kenya's farming sector has been the quality of its research complex. Although African farmers have benefited from access to this research only since their political activity secured it for them, this complex is a colonial and not an *Uhuru* creation. Kenya's research stations were created to meet the needs of the European settlers, who dominated the political life of the colony. The quality of research done on those stations and the sensitivity of their recommendations to the practicalities of commercial agriculture reflected in no small part the power of European farmers who knew what they needed and who had no hesitation in dressing down a research officer whose recommendations had been produced with technical perfection and not profits in mind. Here the "squawk factor," analyzed in chapter 9, worked not only to favor the settlers but also to produce commercially sensitive research.

The outbreak of the Mau Mau rebellion in the late 1940s forced Europeans to make their coffee husbandry technology available to African growers in the early 1950s. With African government imminent, tea and hybrid corn also began to be extended to African farmers. Agricultural research has a long lead period, and it is still too early to tell if the profitable technological innovations produced by European scientists in the colonial and early independence periods will be replicated by the African-dominated research stations of the present. In our view, the results will depend on the atmosphere within which the African scientists and technicians will work and the resources with which they are provided.

The highly favorable climate for technological research that existed in the colonial period has changed subtly but significantly for the worse with the rise of African nationalism. In 1958–1959, when the white settlers were firmly in political power, agricultural and animal husbandry research claimed K£ 194,618 (U.S. $544,930) or 21 percent of the development budget. In 1963–1964, under an African government and on the eve of *Uhuru*, the comparable figure was K£ 222,670 (U.S. $623,476) or 17 percent of the development estimates. By 1970–1971, after six years of independence, development expenditures on agricultural and animal research were K£ 232,361 (U.S. $650,611), which is a scant 6 percent of the development budget.[10] Thus the government's contributions to technological development have remained relatively constant in absolute terms but have failed to keep pace with inflation and have represented a steadily diminishing proportion of total government expenditures. It is clear that the interest of an African-dominated government in agricultural and veterinary research is substantially less than that of a white settler-controlled one. The present government has been more interested in making the benefits of past research available to the African farmers, to whom they were previously denied, than it has been with guaranteeing a continuous flow of technological improvements in the sector. This is not to say that the government has been hostile to research; rather that it has shown neglect because its attentions have been focused on other problems. A good deal of the interest that has been shown is a direct result of the initiatives and contributions of various international aid bodies.

The decline in interest in research at the political level and the consequent lag in resources for it is reflected in the relationships between the research stations and the extension services and farmers. The field officers whom we interviewed for our study do not seem to be making demands on research officers. The African owners of large farms, who usually are only part-time managers, are too busy to visit the research stations, and the smallholders never have felt that they had any rights at such institutions. Thus the research establishments no longer feel the pressure of no-nonsense demands that aggressive white settlers once made of them. This change in the structure of interests surrounding the agricultural research enterprise makes it much more of a low pressure, charitable exercise by academically minded researchers and less of a service provided by right to a group of farmers with powerful and technologically discerning leadership. The shift here has been gradual and subtle, but it seems to us to be significant and to bode a long-term decline in the flow of technological innovations unless it is reversed.

The importance of the interest atmosphere to research is well illustrated by the history of new corn seed development in East Africa. The research breakthrough needed for the creation of African hybrid corn seed was made in Tanzania in the 1950s. Yet Kenyan research stations had hybrid seed ready for actual use in the very early 1960s, years before the Tanzanians got theirs out. The greater speed and practicality of the Kenyan research complex seems to us to have been due to the demand structure created by commercial white farmers.

The Conditions for Organizational Effectiveness

Good research is the prerequisite to agricultural progress, but an extension organization must actually reach the small farmer with the new technologies if development is to occur. Is it possible for the extension services of developing countries to be effective? What are the environmental conditions under which they will come to function well?

Tea development in Kenya presents a different but no less interesting pattern than does corn. The major Kenyan innovation in this area was not in husbandry methods but in organization. Tea is a demanding crop, which must be plucked carefully and regularly and be processed within hours of being picked. Previous experience in Asia had indicated that the discipline and economies of a large plantation are necessary if the logistical and quality problems of the crop are to be solved. A few years before Independence, however, the parastatal Kenya Tea Development Authority (KTDA) was able to design an administrative structure with a combination of controls and incentives that has made small growers even better and more efficient than plantations.[11] Today the KTDA is still a model agricultural development agency.

The success of the KTDA reinforces one of the basic conclusions of this study: Kenya's agricultural extension services are indeed capable of being highly effective instruments of economic development with the material and human resources which they already possess. This statement runs directly contrary to one's first impressions when one goes into the field to observe development agencies at work. There one meets the agricultural officer who complains that he "has had to chase his JAAs from market places, *pombe* [beer] parties, aimless cycling all over the countryside, etc. . . . Our extension workers are often too close to the client system and insufficiently trained to fill suitably the 'specialist role.' "[12] This is the perspective that informs much of the "penetration" school of analysis and its conclusion that the underdeveloped countries lack the money and skilled manpower necessary for effective development administration. But we have shown in this study that material resource inadequacies and human problems such as the motivation, insufficient training, and involvement in traditional society of junior staff just cited are not the real problems. Instead they are the symptoms of organizational structures that either encourage such conditions or fail to correct them.[13] If Kenyan extension administration were reformed in ways such as those we have suggested in this study, the government would be an even more effective and uniform force in promoting agricultural development than it is now.

What then are the environmental conditions under which a public development organization in a country like Kenya will adopt an effective administrative structure? The answer has three components: (1) that the agency's clients have political power; (2) that they demand development; and (3) that the agency's administrative leadership interprets these pressures as requiring an effective organizational structure. At the height of the colonial period, Kenya's white settlers were politically powerful; they demanded an extensive set of services to facilitate the development of their farms; and a quite effective collection of agencies was created for their benefit.[14] First with the Mau Mau uprising and then with *Uhuru*, Kenya's African population emerged from the political wilderness and demanded the benefits of economic development for themselves. The responses of the English colonial administrators were the Swynerton Plan of land consolidation and smallholder coffee production, the Kenya Tea Development Authority, the hybrid corn programs, and so on.

Most of Kenya's successful agricultural development programs, even to the present day, have resulted from the combination of forceful African political demands and the responses of English colonial and postcolonial administrators. (Only in the last few years have African administrators begun to fill the positions in which decisions about organizational structure are actually made in agencies such as the KTDA and the agricultural research services.) The crossing of African demand and European administrative response often produced a particularly vigorous strain of development agency. The ideology of European colonialism and neocolonialism predisposes a European to see underdevelopment as a result of individual and social failings and not as caused by exploitation, political weaknesses, or shortages in material resources.[15] Thus the European can blame African poverty on the Africans and get rich with a clear conscience.

Nonetheless, the rise of Africans to political power made it necessary for the colonial administrators in Kenya to create African development in spite of their European ideological predispositions if they were to survive. Most of them were unwilling or unable to attempt the task and disappeared at Independence. A few of them actually enjoyed the new set of African political pressures, however, and prepared to stay on. These European administrators did not change their view of the causes of African poverty. If they had done so, they would have found themselves advocating the African confiscation of the Kenyan White Highlands, for example, not its purchase at a substantial "fair price." They continued to believe that economic development basically comes from human processes that create more resources than they consume and thus can pay their own way. Since these European administrators saw underdevelopment in social terms, it was natural for them to seek improved methods of social and administrative organization in order to achieve growth. Only a few of these excolonial officials were successful in their search, but the skilled and lucky ones produced some important organizational structures, most notably the KTDA. If Europeans still ran Kenya's development agencies, the KTDA is one of the models on which they would pattern their current organizational structures.

But Kenya's Africans have now replaced Europeans as the administrators and architects of its development agencies. It seems to us that these African decision makers respond to the rural pressures for economic development in a significantly different way.[16] In the view of Africans— to offer another oversimplified characterization—economic development is determined primarily by access to material resources and not by their efficient use. Furthermore, they see the allocation of material resources as being ultimately a political matter. The view that development is primarily a distributional problem pervades the entire political demand structure in Kenya. Farmers want loans, not better farming methods. Parents want more schools and more years of education, not better and more relevant teaching. Coffee growers want to know where the next cooperative factory will be built, not how efficiently it will operate. In the rural areas people will go to great lengths to assure their access to future government resources, even at the cost of the inefficient use of the resources that they do have.[17] As Goran Hyden's study of Kenyan cooperatives has made clear, even the smallest questions of resource allocation attract intense interest in the rural areas, but efficiency problems have almost no constituency.[18] This political demand structure is different from that in Europe and America only in degree, not in kind, however. The important variations arise in the administrators' responses to these demands.

Where the European administrator in Kenya responds to political pressures for development by asking how available resources can be used more efficiently, the African seeks to expand the supply of resources, irrespective of their ultimate cost or of how well they can be used. When faced with a decision of how to allocate resources, Kenya's European administrators tend to put them where they can be most productive, while the local administrator puts them where, for political reasons, he most wants development to occur. In our opinion the African's viewpoint comes much closer to the actual nature of economic development than does the European's, although a complete picture would have to include components from both sides. Nonetheless, the tendency of Africans to see development as based on the allocation of resources rather than their efficient use has unfortunate consequences when they are in top administrative and managerial positions. They are preoccupied with allocative questions and are uninterested in efficiency or effectiveness problems. Consequently Kenya's public organizations are far less efficient in their use of society's already scarce resources than they could be.

There has been a great deal of official Kenyan attention given to problems of administrative reform in the last few years, and this would appear to contradict our assertion of African disinterest in these questions. Closer examination actually lends support to our view. The white technical assistance personnel who wrote most of the 1970-1974 *Development Plan* slipped in the following sentences: "Despite its importance, rather little is known about the effectiveness of agricultural extension. For this reason an evaluation of the extension services will be made early in the Plan period."[19] Once the plan had been adopted these advisers moved to carry out the evaluation. At that point the Ministry of Agriculture realized that it was being threatened by the Ministry of Economic Planning and Development and set up its own evaluation working party. All the senior administrators in the ministry were included in the committee's membership, but a young white adviser was given the chairmanship and a junior African agricultural planner the secretaryship. The ministry's African administrators stopped attending after the first few meetings of the working party but the various Europeans stuck it out.[20] Although the report was de facto a European one, a good many of its recommendations were implemented.

The senior African administrators were not hostile to organizational effectiveness, only uninterested in it. Thus also, to the best of our knowledge, no senior African headquarters official participated in the implementation of the report. Most of the work was done by technical assistance personnel provided by the United States government.

A similar pattern is evident in the important Ndegwa commission report.[21] A high-level commission of inquiry was appointed by the president to examine the salary and personnel structures of the Kenyan Civil Service. It seems likely that the primary purpose of the commission was to legitimize a substantial rise in salaries for public servants and particularly for higher civil servants.[22] The Ford Foundation recognized it as also being a possible starting point for administrative reform and offered the commission staff support. Kenya's elite is not antagonistic to organizational reform as long as it threatens no particular important interests, and so the Ford-provided staff were able to write substantial general arguments for it into the report. With the publication of the report the Ford Foundation has given further technical assistance to help the government implement its reform proposals. It is reasonable to expect that some reform will in fact result from this Ford investment.

Our description of these two reform exercises may give the impression that the Kenya government is being manipulated into reorganization through the American provision of aid personnel. This is not actually the case, for we personally have seen Kenyan administrators block reform actions that they regarded as prejudicial to their particular interests on several occasions. The Kenyan elite allows foreign meddling only in the efficiency problems that it feels it would be beneficial to solve. The Kenyan leaders do not attend to these problems themselves, because doing so would distract them from the questions of allocation that most concern them.

The analysis so far implies a neat division of administrative labor, with Kenyans distributing benefits and Europeans increasing the total amount of benefits through organizational efficiency. But, of course, no such division really exists or is possible. Organizational rationality is difficult without at least senior administrative positions being awarded primarily to those whom past performance shows to be the most capable. And Kenya's politics of resource allocation is centrally concerned with jobs, both as an end and as a means. The use of jobs as benefits for the members of one's clientele is a common part of patronage politics. Less familiar is the distribution of jobs as part of a larger strategy in the politics of allocation. In Kenya one can assume that a decision maker will skew the distribution of benefits he controls toward the members of his own ethnic group. There are limits on how far he will be able to go if he is to retain his position, however, and the members of his group are usually disappointed with his performance on their behalf. But the bias toward one's own is still one of the basic and accepted decision-rules for Kenyan administrators. Consequently, important administrative positions are distributed in Kenya not only as benefits in themselves but also with the expectation that they will lead to still more indirect benefits for the others in the recipient's ethnic group. In one way or another, most high-level organizational reforms are ultimately wrecked on the rock of this aspect of Kenyan administrative behavior.

There are only a very few routes by which administrative reforms can escape destruction by ethnic politics in Kenya. One is by initiating them at the provincial and not the national level. The Civil Service rules of the game are that a senior staff member should almost never be assigned to a field post in his home area. Consequently when one moves out of Nairobi, one finds that the administrative authorities have no personal or group stake in the allocative decisions that they are making for their area. This detachment substantially lessens the intensity of ethnic group politics within the field administration. It also gives the authorities the objectivity needed to accept the consequences of administrative rationality. Several of those who have been interested in promoting organizational or program reforms in Kenya have found genuine interest and support for their ideas at the provincial and district levels.[23]

The other occasion on which Kenyans may give vigorous support to administrative rationality is when the dominant group in a particular agency feels that it has a natural competititve advantage there and that its best prospect for legitimating its position is to use rigorously objective decision criteria, particularly on personnel matters. In such situations, one will find Kenyan administrators even more actively interested in organizational efficiency than are field officials. Thus there are places and times at which administrative reform can be carried out through the initiative of Kenyan managers. But these occur because of administrators' definitions of their atypical situations.

The Political Context of Administration

The focus of Kenyan politics on the ethnic distribution of benefits has multiple negative consequences for administrative efficiency. A good example is provided by educational and manpower planning. It is widely recognized that Kenya's educational system bears little rational relationship to the country's manpower needs. The primary and secondary schools and many parts of the university are geared to the production of graduates in numbers and with qualifications that the nation cannot use.[24] Although it is clearly inefficient to train more people for well-paying types of work than there are jobs for them, no one wants to be eliminated from the competition for the positions or their requisite qualifications. Thus tremendous political pressures exist to expand beyond all reasonable proportions those parts of the educational system that lead to good jobs. Pressures for educational expansion also exist in developed countries and produce inefficiencies there as well. But the structure of politics in Kenya gives these pressures particular force. The organizing principle of Kenyan politics is ethnic-group membership and the ethnic units are led by members of the rising elite, who are those most interested in education. Thus, where in developed countries only some of the politically important interest groups are promoting educational expansion, in Kenya all of them are. The leaders of each of Kenya's ethnic groups feel that failure to maximize educational opportunity in their own area will lead to their eventually being preempted from good jobs and influence by those from other areas.[25]

Educational expansion run riot not only is inefficient in itself but also works to promote inefficiencies in other areas. The employment of young men who feel themselves overqualified for their posts leads to a decline in motivation and administrative performance. The sudden filling of most high-level jobs by men from a few, relatively young age-groups makes it very difficult to offer promotions as an incentive at a later point, with negative consequences for efficiency.

What are the conditions under which politicians and administrators do become concerned with governmental effectiveness and efficiency? We think that such concern comes only when they realize that an agency's resources are limited and that its client groups have more to gain by cooperation than competition in using them. This perception can come from a variety of causes. We have already noted that the European administrator's view of African underdevelopment inclined him to a conflict-free, social-efficiency model of development. In the United States the administrators and client-group leaders concerned with a particular agency frequently cooperate because the interest groups which most oppose them are not served by the agency at all and are seeking to deny it resources rather than trying to influence the allocative policies within it. A pattern similar to this is also evident in Tanzania. That nation's socialists hold a view of underdevelopment similar to the Kenyan African one, that its causes are political and due to historical injustices in the allocation of material resources.[26] But since Tanzania's enemies are seen as being foreign capitalists, the peasants and workers can only pursue development by cooperating in the most efficient possible use of the resources that they do control.[27] Thus in Tanzania governmental, program, and administrative effectiveness enjoys much more complete political support than it does in Kenya. For example, strict educational and manpower planning are realities in Tanzania (in marked contrast to the Kenyan situation) and a significant regionalization of central headquarters staff was achieved in the 1972 decentralization exercise. This is not to say that the current administrative picture in Tanzania is very bright. As Henry Bienen makes clear in his *Tanzania*, its political and administrative capacities were quite weak in the first years after its independence, and further dislocations have occurred as it has attempted to replace capitalist economic institutions. The situation has improved in the last few years, however, and in our view will continue to do so.

The otherwise quite different politics of Tanzania and America do have one common dimension—the political conflict groups to which administrators are largely responding are organized around economic divisions and not ethnic or geographical ones. Such functional specialization in political organizations is consistent with the functional specialization of administrative organization. Thus when political conflict is structured along economic lines, the head of each particular administrative agency tends to be faced with a clientele with common political interests. When conflict is ethnically or geographically based, however, it cuts across society's functional divisions and confronts each specialized agency with a clientele that represents the full range of political conflict. In the first instance, political conflict is largely external to the agency and administrators are able to consider the problems of efficiency within the organization, for its goals are fairly clear. In the second situation, the society's political battles are being fought within the agency, leaving unsettled the goals that it should be pursuing on the most salient issues and overriding the consideration of efficiency issues.

This discussion of the political determinates of a concern for administrative effectiveness casts light on one of the more important propositions about development administration advanced by Fred Riggs. Riggs argues that public bureaucracies become more efficient as their political power declines and that legislatures and competitive parties are the best means for keeping them subservient.[28] The idea that administrators, like the rest of us, need to be pressured to perform at their best is sound. But we find unsatisfactory the notion that a particular set of political institutions is necessary to provide pressure.[29] To see why, we have only to ask ourselves what would happen if Kenya were to have vigorous, competitive political parties of the American type and a stronger parliament than it presently has. The answer is that the public services would be torn apart still further by the distributional politics that are already hampering administrative efficiency. Kenya's higher civil servants may not be very responsive to individual members of parliament, but if anything they are too sensitive to political pressures.[30] Far from dominating their political environment, Kenya's managers seem to be overwhelmed by it.[31]

Riggs' error is that he discusses the problems of administrative control in institutions rather than in political conflict groups. Riggs implicitly assumes a situation in which the public services are dominated by the members of an elite social class. Bureaucratic power in this situation would lead to administrative exploitation of other classes. If, as was the case in Europe, the rising political power of the lower classes were expressed through parties and legislatures, these would in fact be the institutions to force the bureaucracy into public rather than self service. In societies such as Kenya, however, class consciousness is only nascent and the Civil Service is crosscut by the country's political conflict groups rather than being dominated by any one of them. Such a situation thrusts administrators into the role of political representatives in the intragovernmental struggles over the distribution of resources. Kenya thus has a "representative bureaucracy" in a sense much more profound than mere social arithmetic.[32] Riggs's analysis is misleading because he fails to recognize that institutions are not what is important to political power but the patterns of conflict between a society's salient social groups.

Conclusions

Does economic underdevelopment render the state too weak to be an instrument for agricultural development? We have seen that the expansion of the agricultural sector of the economy is more difficult when one is dealing with traditionally oriented smallholders. But these problems are surmountable with sufficient technological and organizational innovativeness. Furthermore, good agricultural research complexes and effective extension services can be created with the human and material resources possessed by countries such as Kenya. We have been led to the conclusion that effectiveness of the state as an agent of agricultural development ultimately depends on the structure of political demands.

The creation and maintenance of effective government agricultural development agencies depend first of all on the political power of small farmers and secondly on the strength of their demands for development. Throughout the world the second of these two conditions is now easily fulfilled. As to the first, during the last twenty years small farmers in Kenya and most other developing countries have greatly increased their political power, although they are still a subordinate group and the truly poor farmers have almost no voice at all. The third political requisite for effectiveness usually is not even partially met, however, and concerns the way that political and administrative authorities respond to the demands for development. Most Kenyan Africans see economic development as deriving from access to scarce resources and not from their productive use. When the development process is understood in this way and social conflict groups are organized on ethnic lines, most political and administrative energies are burned up in the competition for resources and too little attention is given to the technological and organizational innovations that are actually essential to successful agricultural development on a national basis. The African's understanding of the causes of underdevelopment is correct as far as it goes, but his disregard for the efficient use of his existing resources is dysfunctional to his economic development. Given the present world balance of power, no amount of politicking is going to transfer adequate Western wealth to the developing nations. To improve their present world position, poor states must strive for increased self-reliance through the efficient use of the resources they do control.

The African conviction of the allocative character of development and the ethnic group structure of Kenya's political conflict normally drive the elite's attention far from questions of economic efficiency. We noted a few situations in which administrative leaders in Kenya are not involved deeply in distributional questions and give their attention to efficiency problems, but these are exceptional. We feel that in the developing world, probably only in countries such as Tanzania, where the political enemy is foreign capitalism, can domestic political interests cooperate enough to provide consistent support for technological innovation, program effectiveness, and organizational rationality.

A sustained role for the state in accelerating agricultural development depends on the simultaneous solutions of resource, technical, and political problems. It has been encouraging to learn that Kenya's financial and manpower resources are much more adequate than they were once thought to be and that the technical problems of policy, organizational structure, and technology are surmountable. It is depressing to be told that the country's rapid development depends on a type of political structure that is very unlikely in the foreseeable future. If we were to hold that effective development administration depended on the simultaneous provision of all requisites, we would be extremely pessimistic about the ability of the government to contribute to Kenya's growth. We would then favor a laissez-faire approach to the economy or a revolutionary stance toward the state wherever possible.

We believe, however, that there are actually significant imbalances in the process of administrative development. In many sectors, extremely intense political competition would engulf any government programs and rationalized state activity in them is largely unachievable. Other program areas are politically receptive (sometimes accidentally) to organizational efficiency, and in these, solutions to the technical and resource problems of development administration can actually be implemented and produce substantial economic benefits. These proofs of the rewards of efficiency in turn help to create an atmosphere in which organizational and technological innovations are appreciated as an aid to development and become politically somewhat more attractive.

Kenya's agricultural extension services are embedded in a social and political environment that erodes their efficiency and effectiveness. The distribution of extension benefits is substantially less politicized than most areas of government activity, however. One of the purposes of our study has been to demonstrate that the extension services have the capacity to be considerably more effective instruments of agricultural development. A wider understanding of this potential can help to create political support for the reforms needed in the ministries of agriculture throughout the developing world.

Notes

Preface

1. René Dumont, *False Start in Africa* (London: Sphere Books, 1968), p. 30.
2. Uma Lele, *The Design of Rural Development: Lessons from Africa* (Baltimore: The Johns Hopkins University Press, 1975), pp. 183–84.
3. Colin Leys, *Underdevelopment in Kenya: The Political Economy of Neo-Colonialism* (Berkeley: University of California Press, 1974), pp. 113–14.
4. Christopher Trapman, *Changes in Administrative Structures: A Case Study of Kenyan Agricultural Development* (London: Overseas Development Institute, 1974), pp. 19, 77. Lele, *The Design of Rural Development*, p. 67.
5. Trapman, *Changes in Administrative Structures*, pp. 85–86.
6. Institute for Development Studies, "The Second Overall Evaluation of the Special Rural Development Programme," mimeographed (Nairobi: Institute for Development Studies, University of Nairobi, 1975), pp. 2.18–19, 8.14. Joseph Ascroft, Niels Röling, Joseph Kariuki, and Fred Chege, *Extension and the Forgotten Farmer: First Report of a Field Experiment* (Wageningen: Afdelingen voor Sociale Wetenschappen aan de Landbouwhogeschool, 1973).

Chapter 1

1. René Dumont, *False Start in Africa* (London: Sphere Books, 1968), p. 165.
2. Henry Bienen, *Tanzania: Party Transformation and Economic Development*, expanded edition (Princeton: Princeton University Press, 1970), esp. p. 449.
3. Kenya, *Report of the Commission of Inquiry (Public Service and Rumuneration Commission) 1970–71*, D. N. Ndegwa, Chairman (Nairobi: Government Printer, 1971), p. 85.
4. Robert Jackson, unpublished working paper, Nairobi, 1970.
5. E.g., L. Cliffe, J. S. Coleman, and M. R. Doornbos, eds., *Government and Rural Development in East Africa* (The Hague: Martinus Nijhoff, 1976).
6. Kenya, Ministry of Agriculture, "Final Report of the Working Party on Agricultural Extension Services," mimeographed (Nairobi: 1970), pp. 6–7.
7. Kenya, *Development Plan 1970–74* (Nairobi: Government Printer, 1969), pp. 2, 11.
8. Y. P. Ghai and J. P. W. B. McAuslan, *Public Law and Political Change in Kenya* (Nairobi: Oxford University Press, 1970), pp. 79–124, 274–86, 291–300. M. P. K. Sorrenson, *Origins of European Settlement in Kenya* (Nairobi: Oxford University Press, 1968), pp. 150–51.
9. All prices given in this book were current in the early 1970s when our research was conducted. The exchange rates to U.S. dollars are also those prevailing at that time.

10. R. G. Lever, *Agricultural Extension in Botswana* (Reading, U.K.: University of Reading, Department of Agricultural Economics, 1970), pp. 102, 103, 106.

11. R. G. Saylor, "A Social Cost/Benefit Analysis of the Agricultural Extension and Research Services in Selected Cotton Growing Areas of Western Tanzania" (paper presented at the Universities of East Africa Social Science Conference, Dar es Salaam, December 27–31, 1970).

12. R. G. Saylor, "Variations in Sukumaland Cotton Yields and the Extension Service" (paper presented at the East African Agricultural Economics Conference, Dar es Salaam, April 1–4, 1970), pp. 29, 35.

13. There are a large number of such successful experiments. Two of those for which the results have been published are: Joseph Ascroft, Niels Röling, Joseph Kariuki, and Fred Chege, *Extension and the Forgotten Farmer: First Report of a Field Experiment* (Wageningen: Afdelingen voor Sociale Wetenschappen aan de Landbouwhogeschool, 1973) and B. D. Shastry, "Quickening the Pace of Agricultural Development," *Economic and Political Weekly* (Bombay) 6 (June 1971): A95–98.

14. E.g. Herbert F. Lionberger and H. C. Chang, *Farm Information for Modernizing Agriculture: The Taiwan System* (New York: Praeger, 1970); David W. Kidd, "A Systems Approach to Analysis of the Agricultural Extension Service of Western Nigeria" (Ph.D. dissertation, Department of Extension Education, University of Wisconsin, 1971); Radha C. Rao, "Communication Linkages in Transfer of Agricultural Technology," *Economic and Political Weekly* (Bombay) 7 (December 1972): A157–70.

15. Jon R. Moris, "Managerial Structures and Plan Implementation in Colonial and Modern Agricultural Extension," in *Rural Administration in Kenya*, ed. by D. K. Leonard (Nairobi: East African Literature Bureau, 1973), p. 100.

16. Robert Chambers, *Settlement Schemes in Tropical Africa* (London: Routledge and Kegan Paul, 1969), p. 145.

17. This description portrays the standard message model of agricultural extension. Philip M. Mbithi, "Agricultural Extension as an Intervention Strategy," in *Rural Administration in Kenya*, ed. by Leonard, pp. 79–80.

18. These data are derived from a preliminary investigation done by the author before launching the major survey reported in this study. For details of the sample and questionnaire, see David K. Leonard, "Some Hypotheses Concerning the Organization of Communication in Agricultural Extension," mimeographed (Nairobi: Institute for Development Studies, University of Nairobi, 1970).

19. Kenya, Ministry of Agriculture, "Final Report of the Working Party on Agricultural Extension Services," pp. 6–7.

20. We found a mimeographed advertisement for recruitment to this cadre on a district headquarters notice board after being told by Nairobi officers that there were to be no new JAAs. The ministry probably allows additional recruitment in "special cases."

21. Kenya, Ministry of Agriculture, "Final Report of the Working Party on Agricultural Extension Services," p. 8.

22. Extension agents in the U.S.A. all hold degrees; those in Holland, for example, have diplomas. Private communication, Niels Röling, extension specialist.

23. For Nigeria, see R. K. Harrison, "Work and Motivation: A Study of Village-Level Agricultural Extension Workers in the Western State of Nigeria," mimeographed (Ibadan: Nigerian Institute of Social and Economic Research, 1969), chap. 6.

24. This name and those of the other extension staff mentioned in this book have been changed to keep the promise of confidentiality we made to them.

25. A 1968 survey of Vihiga Division found that 25 percent of the farms had houses with *mabati* roofs. See J. Heyer and J. Ascroft, "The Adoption of Modern Practices on Farms in Kenya: Preliminary Results of a 1968 Survey of Farms Across Kenya" (paper presented at the Universities of East Africa Social Science Conference, Dar es Salaam, December 27–31, 1970), p. 11.

26. This important but undefined concept appears in Kenya, Ministry of Agriculture, "Final Report of the Working Party on Agricultural Extension Services," p. 13.

27. Kenya, *Report of the Commission of Inquiry*, D. N. Ndegwa, chairman, pp. 81, 91.

28. Michel Crozier, *The Bureaucratic Phenomenon* (Chicago: The University of Chicago Press, 1964), pp. 210–36.

29. P. M. Blau and W. R. Scott, *Formal Organizations* (San Francisco: Chandler Publishing Company, 1962), pp. 199–202.

30. Goran Hyden, *Efficiency versus Distribution in East African Cooperatives: A Study in Organizational Conflicts* (Nairobi: East African Literature Bureau, 1973), pp. 207, 208, 223.

31. C. T. Leys and P. Stamp, "Organization and Development: Dilemmas of Administrative Training in Kenya," mimeographed (Nairobi: University of Nairobi, Department of Government, 1971).

32. Silas M. Ita, "The Changing Role Expectations of the Chiefs in Mbere Division 1900–1971," mimeographed (Nairobi: University of Nairobi, Department of Government, 1972).

33. For example, see Blau and Scott, *Formal Organizations*, and its extensive bibliography.

34. Amitai Etzioni, *A Comparative Analysis of Complex Organizations* (New York: Free Press, 1961), pp. 12, 23.

35. We are aware of only one—R. K. Harrison's "Work and Motivation."

36. Amitai Etzioni, *Modern Organizations* (Englewood Cliffs, N.J.: Prentice-Hall; 1964), p. 41.

37. Julius K. Nyerere, *Freedom and Socialism: Uhuru na Umoja* (Dar es Salaam: Oxford University Press, 1968), pp. 344–45.

38. The author was a member of the committee which produced Kenya, Ministry of Agriculture, "Final Report of the Working Party on Agricultural Extension Services." See also our "Organizational Structures for Productivity in Agricultural Extension," in *Rural Administration in Kenya*, ed. by Leonard, e.g., p. 153.

Chapter 2

1. Uma Lele, *The Design of Rural Development: Lessons from Africa* (Baltimore: Johns Hopkins University Press, 1975), p. 62.

2. For example, Easton assigns administrative agencies and their performances to the output side of systems analysis. David Easton, *A Systems Analysis of Political Life* (New York: John Wiley and Sons, 1965), pp. 355–56, 374.

3. W. F. Ilchman, "The Unproductive Study of Productivity," *Comparative Political Studies* 1, no. 1 (1968): 227–49.

4. Kenya, Ministry of Agriculture, "Final Report of the Working Party on Agricultural Extension Services," mimeographed (Nairobi: 1970), p. 2.

5. For a full discussion of factors which should be taken into account in making such methodological decisions, see D. K. Leonard and K. Prewitt, "Quantification, Productivity and Groups," in *Developing Research on African Administration: Some Methodological Issues*, ed. by Adebayo Adedeji and Goran Hyden (Nairobi: East African Literature Bureau, 1974), pp. 73–108.

6. E.g. R. G. Saylor, "A Social Cost/Benefit Analysis of the Agricultural Extension and Research Services in Selected Cotton Growing Areas of Western Tanzania." (paper presented at the Universities of East Africa Social Science Conference, Dar es Salaam, April 1–4, 1970).

7. D. K. Leonard, "A Proposal for an Evaluation of the Agricultural Extension Services in Kenya," mimeographed (Nairobi: Institute for Development Studies, University of Nairobi, 1970).

8. E.g. David K. Leonard with Bernard Chahilu and Jack Tumwa, "Some Hypotheses Concerning the Impact of Kenya Government Agricultural Extension on Small Farmers," mimeographed (Nairobi: Institute for Development Studies, University of Nairobi, 1970).

9. For example, Agricultural Informedness and Visit Effort (explained below) are uncorrelated ($r = -.08$; N = 169). The number of staff meetings held in a location is positively correlated with an agent's Agricultural Informedness ($r = .29$; N = 169; $p < .01$) and is negatively correlated with his Visit Effort ($r = -.20$; N = 169; $p < .02$).

10. Although the ministry also recommends some innovations unlikely to be profitable under some ecological conditions or when farm labor is costed at its full value, following these recommendations would only rarely be damaging to the farmer's interests. Certainly the great bulk of the ministry's innovations are economically desirable for the small farmer.

11. E. R. Watts, "Agricultural Extension in Embu District of Kenya," *East African Journal of Rural Development* 2 (1969): 69–71; Leonard with Chahilu and Tumwa, "Some Hypotheses Concerning the Impact of Kenya Government Agricultural Extension on Small Farmers," pp. 6–7.

12. Leonard with Chahilu and Tumwa, "Some Hypotheses Concerning the Impact of Kenya Government Agricultural Extension on Small Farmers," pp. 11–12.

13. The series on corn fertilizer was selected because of the importance of hybrid corn throughout the province's extension program. We consulted Alistair Allen of the Maize [Corn] Research Station at Kitale first and then based the questions and their answers on the pamphlet circulated by the Kenya Seed Company with its hybrid corn seed.

There was a major program to expand grade-cattle ownership and milk production in Western Province at the time of our research. The questions on milk cow feeding were developed after consultations with John Peberdy, then head of the Animal Production Division of the Ministry of Agriculture. We also consulted his assistants and read the relevant research station findings. About two-thirds through our interviews, we found that the provincial Animal Husbandry Officer had just received approval for a set of recommendations slightly different from those with which we were working. It was too late to adjust our questionnaire but, as he had not yet begun to diffuse the new recommendations we do not believe that our results were seriously affected.

East coast fever is the most common cattle disease in Kenya. The source of our information on this set of questions was Dr. Gibson, then head of the Animal Health section at the Animal Health and Industry Training Institute. He is a former District Veterinary Officer with long field experience in Kenya. We have encountered some Veterinary Officers who would add to the list of symptoms that Dr. Gibson gave us. Although it is unfortunate that our scoring may not have been complete, the problem is minor, because we gave points for symptoms correctly mentioned and deducted nothing for wrong answers.

Our questions on coffee were suggested to us by Mr. Wallis, a former Senior Coffee Officer. The information tested is that published by the Coffee Research Station at Ruiru. Nonetheless, the color of a coffee bean infected by the berry borer may vary from that indicated, as we ourselves saw in the field. Again, we believe the resulting bias to be minor.

Our series of questions on cotton were suggested by the cotton entomologist at the National Agricultural Research Laboratories in Nairobi. The information tested is drawn from published material. We have encountered some disagreement whether the spiny bollworm, as opposed to the American one, is a major problem in Western Province. Nonetheless, the ability of agents to answer these questions was reasonable and is well correlated with ability on other technical questions asked.

Finally, we are grateful to Joseph Ascroft, then of the Institute for Development Studies, University of Nairobi, for suggestions on the precoding and scoring of the questions we used.

Further details, including distributions, can be found in D. K. Leonard, et al., "The Work Performance of Junior Agricultural Extension Staff in Western Province: Basic Tables," mimeographed (Nairobi: Institute for Development Studies, University of Nairobi, 1971).

14. David K. Leonard, "Some Hypotheses Concerning the Organization of Communication in Agricultural Extension," mimeographed (Nairobi: Institute for Development Studies, University of Nairobi, 1970), pp. 11–13.

15. It may be objected that not all farmer visits will be of equal length. We anticipated this problem and asked our respondents about the duration of each visit made. Sixty percent of the visits were short. We have experimented with an alternative indicator of Visit Effort that treats one long visit as equal to two short visits. Although we feel that this alternative is superior in principle, we find that the results of correlation analysis with the two are very similar. The limitations of the computer facilities available to us in Nairobi also would have made it very difficult to use the alternate version for all our analyses. In the circumstances we have decided to work with the less ideal indicator.

16. Unfortunately, our three interviewers did not choose the respondents with whom they worked randomly. Edwin Luchemo, the interviewer who received only honest responses, worked exclusively with staff in Kakamega District. Jack Tumwa worked exclusively in Bungoma and Busia Districts, and Humphries W'Opindi covered the entire area. The Visit Effort means of Tumwa and W'Opindi in Busia and Bungoma are quite similar, whereas those of Luchemo and W'Opindi in Kakamega are significantly different. We therefore took the proportionate difference between Luchemo's and W'Opindi's scores in Kakamega and reduced all of W'Opindi's and Tumwa's scores by that proportion.

17. Rural Development Research Committee, "An Interim Report on the Evaluation of Agricultural Extension," Rural Development Paper No. 5, mimeographed (Dar es Salaam: University College of Dar es Salaam, 1968).

18. Kenya, Ministry of Agriculture, "Final Report of the Working Party on Agricultural Extension Services," p. 2.

19. Kenya, Ministry of Finance and Planning, Statistics Division, *Kenya Population Census, 1969*, 1 (Nairobi: Government Printer, 1970): 67 (hereafter cited as *Census* 1).

20. Ibid.; Ministry of Finance and Planning, Statistics Division, *Kenya Population Census, 1969*, 3:52 (hereafter cited as *Census* 3).

21. East African Meteorological Department, *Monthly and Annual Rainfall in Kenya during the 30 Years 1931 to 1960* (Nairobi: Meteorological Department of the East African Common Services Organization, 1966), p. 13 (hereafter cited as *Rainfall in Kenya*).

22. J. Heyer and J. Ascroft, "The Adoption of Modern Practices on Farms in Kenya " (paper presented at the Universities of East Africa Social Science Conference, Dar es Salaam, December 27-31, 1970), pp. 4, 9.

23. *Census* 1:64, 3:51.

24. *Rainfall in Kenya*, p. 46.

25. Heyer and Ascroft, "The Adoption of Modern Practices on Farms in Kenya," pp. 4, 9.

26. *Census* 1:63.

27. *Census* 3:50.

28. *Rainfall in Kenya*, pp. 4, 15.

29. There were eleven at the time we did our research, but one more has recently been created.

30. The pyramid of administrative units in Kenya is: nation, province, district, division, location, sublocation, and village. A "village" is actually a widely dispersed set of farms and not a group of houses.

31. Leonard, "Some Hypotheses Concerning the Organization of Communication in Agricultural Extension"; Leonard with Chahilu and Tumwa, "Some Hypotheses Concerning the Impact of Kenya Government Agricultural Extension on Small Farmers."

32. We are grateful to H. C. A. Somerset of the Institute for Development Studies, University of Nairobi, for having made this point clear to us when we were making our sampling decisions.

33. The full list of sampled locations is Bungoma District: Elgon, Kimilili, East Bukusu, South Malakisi; Busia District: Bukhayo, Marach, Bunyala; Kakamega District: Idakho, Bunyala, North Kabras, Wanga Mukulu, South Wanga, Tiriki, and East Bunyore. North and South Maragoli and West Bunyore were excluded from our sample because they had been the sites of our pretest.

34. These five are Bungoma District: Elgon and Kimilili Locations; Busia District: Bukhayo Location; Kakamega District: Idakho and South Wanga Locations.

Chapter 3

1. Frederick W. Taylor, *The Principles of Scientific Management* (New York: Harper and Brothers, 1923), pp. 9, 10, 121.

2. F. J. Roethlisberger and William J. Dickson, *Management and the Worker* (Cambridge, Mass.: Harvard University Press, 1939).

3. The Hawthorne studies are summarized, among other places, in P. M. Blau and W. R. Scott, *Formal Organizations* (San Francisco: Chandler Publishing Company, 1962), pp. 89-93.

4. Reinhard Bendix, *Work and Authority in Industry*, rev. ed. (New York: Harper and Row, 1963), pp. 312-17.

5. P. M. Blau and W. R. Scott, *Formal Organizations* (San Francisco: Chandler Publishing Company, 1962), p. 87.

6. Bendix, *Work and Authority in Industry*, pp. 312-17.

7. Ibid., p. 265.

8. Giovanni Arrighi and John S. Saul, "Socialism and Economic Development in Tropical Africa," *The Journal of Modern African Studies* 6 (August 1968): 156.

9. Colin Leys, "Politics in Kenya: The Development of Peasant Society," mimeographed (Nairobi: Institute for Development Studies, University of Nairobi, 1970), pp. 6, 7, 20.

10. Ibid., p. 21.

11. Goran Hyden and Colin Leys, "Elections and Politics in Single-Party Systems: The Case of Kenya and Tanzania," *British Journal of Political Science* 2, no. 4 (1972): 261-92; Goran Hyden, *Efficiency versus Distribution in East African Cooperatives* (Nairobi: East African Literature Bureau, 1973).

12. Ralf Dahrendorf, *Class and Conflict in Industrial Society* (Stanford, Calif.: Stanford University Press, 1959), p. 138 (hereafter cited as *Class*.)

13. Ibid., p. 165.

14. Ibid., pp. 237-38.

15. Dahrendorf solves this problem in a different way. Ibid., p. 126.

16. Ibid., pp. 178-82.

17. These observations are modifications of those in Kenneth Prewitt, "The Functional Justification of Inequality and the Ndegwa Report: Shaping an Ideology" (paper presented at the East African Universities Social Science Council Conference, Nairobi, December 19-23, 1972), pp. 2-7. Issac Balbus, "Ruling Elite Theory vs. Marxist Class Analysis," *The Monthly Review* 23 (May 1971): 39.

18. Some indirect evidence from the United States military, particularly on recruitment and social intercourse, can be found in Morris Janowitz, *The Professional Soldier* (New York: Free Press, 1960), pp. 54-102, 196-204.

19. Karl Marx and Friedrich Engles, *Basic Writings on Politics and Philosophy*, ed. by Lewis S. Feuer (Garden City, N.Y.: Doubleday, 1959), pp. 338-39 (hereafter cited as *Basic Writings*.)

20. J. A. Schumpeter, "Die sozialen Klassen im ethnisch homogenen Milieu," in *Aufsatze zur Soziologie* (Tubingen, 1953), p. 152, quoted in Dahrendorf, *Class*, p. 191.

21. Dahrendorf, *Class*, p. 187.

22. Ibid., p. 188.

23. Marx and Engels, *Basic Writings*, pp. 338-39.

24. Dahrendorf, *Class*, p. 239.

25. Blau and Scott, *Formal Organizations*, pp. 160-62.

26. Dahrendorf, *Class*, pp. 166-67, 170-71.

27. A fuller discussion of this point is provided by John J. Okumu, "The Socio-Political Setting," in *Development Administration: The Kenyan Experience*, ed. by G. Hyden, R. Jackson, and J. Okumu (Nairobi: Oxford University Press, 1970), pp. 36–39.

28. The Animal Health (or Veterinary) personnel used to have a separate organization from the Agriculture staff. For this reason and others internal to our study, we analyzed the Animal Health and Agriculture junior staff interviews separately. We have examined the patterns in both sets of data on the questions discussed in this chapter, and they are basically the same. Rather than complicate the following presentation with two parallel sets of data, we will analyze only the Agriculture junior staff interviews. This gives us a sample of 169, against 44 for Animal Health personnel. The same constraints did not apply to the analysis of the senior staff data and so the six veterinary personnel are included in those figures.

29. For similar data about civil servant social isolation in rural Tanzania, see H. U. E. Thoden van Velzen, "Staff, Kulaks and Peasant," in *Socialism in Tanzania*, vol. 2, *Policies*, ed. by Lionel Cliffe and John S. Saul (Nairobi: East African Publishing House, 1973), pp. 153–79.

30. The comparison with senior staff is obvious, as officers work outside their home areas and retain friends in both places. The assertion about the average small farmer is based upon impressions only.

31. My Luhya research assistants and I established these categories on the basis of our perceptions of status differentia in Western Province. They are judgmental only and open to criticism, even though we believe them to be basically accurate. The main problem in the classification system is where to place farmers who are not running large commercial enterprises. Generally, the progressive farmers would have been put in the Lower Elite category and the others in the Nonelite one. Unfortunately, there were doubtless errors of judgment when the coding was done in the interview. We believe that this problem is not serious enough to invalidate the results.

32. We use the term *progressive farmer* for someone who has adopted several farm innovations and is growing, at least in part, for the cash market. Depending on the exact definition, these constitute 15 percent or less of the farmers in Western Province. This point is discussed in detail in chapter 9.

33. Thoden van Velzen, "Staff, Kulaks and Peasant."

34. Okumu, "The Socio-Political Setting," pp. 34–42.

35. Marx and Engels, *Basic Writings*, p. 338.

36. Schumpeter, quoted in Dahrendorf, *Class*, p. 191.

37. The category *supervisory AAs* contains all the first-line supervisors and a few other AAs who were working on specialist duties at divisional level directly under the AAO but with no staff under them. Their status was clearly equivalent to that of the AA supervisors.

38. Blau and Scott, *Formal Organizations*, p. 122.

39. These differences cannot be proved to be statistically significant because of the small numbers of cases in which the AAO is the immediate supervisor. The tests for differences between proportions show a 15 percent probability that the crucial differences could have occurred by chance alone. We have decided that in this case the greater danger is the probability of a Type II (or Beta) error if we did not offer this analysis. Helen M. Walker and Joseph Lev, *Statistical Inference* (New York: Holt, Rinehart and Winston, 1953), pp. 77–79, 60–61.

40. With tribe the same and class different, we have 2.41 – 1.45 = .96. With class the same and tribe different, we have 1.45 – .36 = 1.09.

41. We might note that where both quasi groups belong to the same tribe in Kenya, clan or other kinship differences would generally take the place of tribe in group cleavages. Thus the same sort of analysis usually can be made in that type of case. Hyden and Gobanga both discuss cases in which intratribal kinship affiliations override interest group as a basis for intraorganizational conflict. Nonetheless, we have evidence of other Kenyan situations in which clan and ethnicity do not override interest group as the basis of organizational cleavage. Mukuha found in a Nairobi bank that hierarchical pressures had been sufficient to create interest group unity among normally antagonistic racial groups. Muigai and Njooro report that interest group completely overrode locality-kinship affiliations in an all-Kikuyu organization. We will reflect further on the larger issue of the bases of organizational cleavage in Kenya in chapter 11. Hyden, *Efficiency versus Distribution in East African Cooperatives*, pp. 207–13; I. Gobanga, "Hierarchical Independence, Subordinate Loyalty, and Power in a Kenyan Rural Bank," J. N. Mukuha, "Power Relations in a Nairobi Bank," and D. Muigai and S. Njooro, "Authority and Class Consciousness in the Nyeri Ministry of Works Mechanical Workshop," all three in "Case Studies in Administrative Behaviour," ed. by D. K. Leonard, mimeographed (Nairobi: Department of Government, University of Nairobi, 1973).

42. Dahrendorf, *Class*, p. 187.

43. An analysis of primary exam performance in Uganda found the rural poor performing better than the rural well-to-do, an unstable relationship if employment is the ultimate source of most wealth. H. C. A. Somerset, private interviews, Nairobi, Kenya, 1969, 1972.

44. Peter Marris and Anthony Somerset, *The African Businessmen* (London: Routledge and Kegan Paul, 1971), p. 66.

45. Prewitt, "The Functional Justification of Inequality and the Ndegwa Report."

46. Our current research on the social origins of students at the University of Dar es Salaam strongly supports these points. See also V. Subramaniam, "The Social Background of Zambia's Higher Civil Servants and Undergraduates," in "Proceedings of the Universities of East Africa Social Science Council 5th Annual Conference" (Nairobi: University of Nairobi, 1969), p. 1683. International Labour Office, *Employment, Incomes and Equality: A Strategy for Increasing Productive Employment in Kenya* (Geneva, 1972), pp. 522–23.

47. Marx and Engels, *Basic Writings*, pp. 338–39.

48. As our sampling unit was the location and not the division, we only rarely interviewed all the staff in a division. This means that some staff may have had friends who would have named them but were not interviewed.

49. Fred N. Kerlinger, *Foundations of Behavioral Research* (New York: Holt, Rinehart and Winston, 1964), p. 559.

50. It might be objected that it is much more difficult to be friends with all the members of a large group than of a small group and that a measure of in-group friendship choices should take this into account. Group size certainly does have an effect. If the junior staff were to name the same number of friends within groups of five that they now do for ones averaging ten, the scores on our cohesiveness measures would jump considerably. The proportion of reciprocal friendship choices made out of those possible would be .28 and the average proportion of other group members named as

friends would be .47. (Even these figures would not be particularly high for groups of five, however, and would still indicate that real social cohesion was rare.) Should we then make an adjustment for group size in our measurements of cohesion? The answer depends on what the measure is supposed to indicate. If we want to judge the degree of positive feeling of members toward their groups, then we would adjust for size. Five friendship choices in a group of ten would indicate much the same liking for the group as would five choices in a group of twenty. If, however, we want to measure the social cohesion of a group, particularly as it might affect the group's capability for collective action, we would not want to make adjustments for group size. Large groups do have more difficulty achieving high scores on our friendship measures and they also have more difficulty in achieving social cohesion. Since we want to know about cohesion and not positive feelings toward the group, the measures given in the body of the text are the appropriate ones. When we use cohesiveness as an explanatory variable for group behavior (as we do in chapter 4), we do so because it is better related to that behavior than are alternative measures that indicate the positive feelings of the members toward their groups. (Other such measures tried were the proportion of all friends named who work in the ministry and the proportion of all ministry friends named who are in the respondent's work group.)

51. Gaetano Mosca, *The Ruling Class: Elementi di Scienza Politica*, ed. by Arthur Livingston, trans. by Hannah D. Kahn (New York: McGraw-Hill, 1939), p. 53.

52. The measure made here is Coleman's *h*. (James S. Coleman, "Relational Analysis: The Study of Social Organizations with Survey Methods," *Human Organization* 17, no. 4 [Winter 1958-59]: 34–36.) Coleman's method of calculating this measure does not apply perfectly to our case, and we had to make some simplifying assumptions in order to use it. (Of course we did keep its statistical meaning and range from + 1.00 to -1.00, with 0.00 indicating no bias.) Coleman assumes that one knows the relative size of each group of individuals who are available for choice. In our case we did not, for we had allowed the respondents to name any friends they had in the province. Technically this would have made the population of the whole area available for choice, but to have proceeded with such an assumption would have badly biased our analysis. The overwhelming majority in the province are Luhya small farmers. As was demonstrated in table 8, all senior staff discriminate against poorer people in their choice of friends. This tendency automatically would have made it appear that the non-Luhya staff had a tribal bias when in fact they might only have a class one. To control for this class factor, we somehow had to estimate the proportion of those of the same status as the senior staff who belong to each tribe. Our data gave us two possibilities. The first was to use the senior staff themselves as a sample of the tribal distribution of those in their social circles. The second was to use the total pool of their friendship choices as such a sample. As it happens, both methods produce very similar results in the overall calculation of Coleman's *h*. We have used the first because it indicates slightly less tribalism (+ .19) than does the second (+ .21), and no tribalism is the null hypothesis.

Chapter 4

1. P. M. Blau and W. R. Scott, *Formal Organizations* (San Francisco: Chandler Publishing Company, 1962), pp. 92–96.

2. E.g. Donald Roy, "Quota Restriction and Goldbricking in a Machine Shop," *American Journal of Sociology* 57 (1952): 427–42.

3. Other than the existence of group values, the only hypotheses we could think of to explain these differences were variations in farm densities or in amounts of information to be conveyed. Neither is tenable when tested against the data.

4. In a way it is remarkable that any of the junior staff gave honest answers in the whole series of questions reported in this chapter. We think that we got the cooperation we did because we had already conducted one set of interviews revealing some clear junior staff weaknesses but had, as promised, kept these from reaching any superior.

5. Stanley E. Seashore, *Group Cohesiveness in the Industrial Work Group* (Ann Arbor: Institute for Social Research, University of Michigan, 1954), pp. 63-80.

6. This relationship does not result from any reduction in the amount of time available for farm visits due to meetings. Visit Effort is an average of what is done in the blocs of time devoted solely to farm visits: the measure is completely independent of the amount of time spent in meetings or in other duties.

7. The rate is $N(N-1)/2$, when N = number in the group. V. A. Graicunas, "Relationship in Organization," in *Papers on the Science of Administration*, ed. by L. Gulick and L. Urwick (New York: Institute of Public Administration, 1937), pp. 183-87.

8. P. M. Blau, "A Formal Theory of Differentiation in Organizations," *American Sociological Review* 25, no. 2 (April 1970): 201-18.

9. Regression analysis theoretically demands the normal distribution of all variables. Except in small samples, however, it is fairly resistant to problems of nonnormality. Nonetheless, in our group analyses we are dealing with small samples and therefore have checked all our variables for normality. Following the advice of James Morgan of the Survey Research Center, University of Michigan, we have achieved approximate normal distributions by moving any extreme values to within two standard deviations of the mean and making the other adjustments necessary to preserve the original rank ordering of the data.

10. We first discovered this effect through a visual inspection of our group data. To measure and test it we used the following procedure. The mean group size is 10.2 and the median is 10. From the number in each group we subtracted a constant of 9.9, giving us a positive number for all groups of medium and large size and a negative one for all small ones. This modified size variable was then multiplied by the friendship cohesion one (the average proportion of other group members named as a friend). The resulting variable gives us a measure of the magnitude of two interacting variables, but with negative values for the small groups. This Size-Cohesion Interaction Effect was then correlated with Visit Effort ($r = .75$). We then tested to see whether this composite variable had any statistically significant explanatory power apart from size (the more powerful of its two components). The partial correlation coefficient of Visit Effort and Size-Cohesion Interaction Effect with size controlled is $r = .50$, which is significant at the .05 level.

11. Blau and Scott, *Formal Organizations*, p. 95.

12. Milton L. Blum and James C. Naylor, *Industrial Psychology: Its Theoretical and Social Foundations* (London: Harper and Row, 1968), pp. 373-75; James G. March and Herbert A. Simon, *Organizations* (New York: John Wiley and Sons, 1958), p. 48.

13. Blau and Scott, *Formal Organizations*, p. 96.

14. Ibid., pp. 129-30, 179-83; and Michel Crozier, *The Bureaucratic Phenomenon* (Chicago: The University of Chicago Press, 1964), pp. 28-29, 109-10.

15. Blau and Scott, *Formal Organizations*, p. 48. Our interpretation here is based on

the relationship reported but argues that the direction of causation may be the opposite of that offered by the authors.

16. Support for this assumption can be found in G. C. Homans, *The Human Group* (London: Routledge and Kegan Paul, 1951), p. 141.

17. The effects of both status and Explanatory Ability persist even when one controls for the other. The correlation between Popularity and LAA status with Ability controlled is $r = .28$. That between Popularity and Ability with LAA status controlled is $r = .14$, $p < .05$.

18. Blau and Scott, *Formal Organizations*, p. 122.

19. This suggestion was made to us by A. Paul Hare of Haverford College.

20. These three items will also stand up to multiple regression analysis, so they stand as independent pieces of evidence that goal achievement leads to popularity. One information item (about grade cattle feeding) is negatively correlated with popularity. We believe this to be an artifact, however, as knowledge about grade cattle is highest in the areas where friendship cohesion happens to be lowest.

21. Blau and Scott, *Formal Organizations*, p. 95.

22. These facts were ascertained through the following formula:

$$D_{ij} = \frac{X_{ij} - \overline{X}_i}{S_i}$$

where X_{ij} = Visit Effort of the most popular subordinate agent (or the LAA) in the ith group, $\overline{X}_i$ = the average of all Visit Effort scores in the ith group, S_i = the standard deviation of all Visit Effort scores in the ith group, and D_{ij} = the standardized deviation of the Visit Effort of the most popular subordinate agent (or the LAA) from the mean Visit Effort of his group.

If the individuals whose scores are so measured are deviating from their group means in exactly the same way as the other agents, the mean of all D_{ij}s will equal zero ($\overline{X}_D = 0$) and the standard deviation will be one ($S_D = 1$). When the computations are made we find that, for the most popular subordinate agents, $\overline{X}_D = 0.28$ and $S_D = 0.99$. For the LAAs, $\overline{X}_D = -0.105$ and $S_D = 1.142$. None of these differences is statistically significant. The analysis described here was made in an essay by R. M. K. Kilavuka, while he was studying with us at the University of Nairobi.

23. Blau and Scott, *Formal Organizations*, pp. 104–6.

24. J. Okigo and G. A. Owuor, "The Nairobi Phonogram Section of the East African Posts and Telecommunications," in "Case Studies in Administrative Behaviour," ed. D. K. Leonard, mimeographed (Nairobi: Department of Government, University of Nairobi, 1973).

Chapter 5

1. See Y. P. Ghai and J. P. W. B. McAuslan, *Public Law and Political Change in Kenya* (Nairobi: Oxford University Press, 1970), pp. 300–304; Silas M. Ita, "The Changing Role Expectations of the Chiefs in Mbere Division 1900–1971," mimeographed (Nairobi: University of Nairobi, Department of Government, 1972).

2. For example, the 42 final year students in my 1973 Administrative Behaviour class at the University of Nairobi were adamant that authoritarian supervision is the most

effective in Kenya—until they had done their own research on the topic. See also Muigai and Njooro, "Authority and Class Consciousness in the Nyeri Ministry of Works Mechanical Workshop," in "Case Studies in Administrative Behaviour," ed. D. K. Leonard, mimeographed (Nairobi: Department of Government, University of Nairobi, 1973), pp. 26.5, 26.8.

3. Reinhard Bendix, *Work and Authority in Industry*, rev. ed. (New York: Harper and Row, 1963), pp. 288–89, 294–95.

4. We are aware of the traps involved in using multiple regression analysis on social data and do not believe that we have committed any methodological errors here. For a discussion of the hazards, see Robert A. Gordon, "Issues in Multiple Regression," *The American Journal of Sociology* 73, no. 5 (March 1968): 592–616.

5. This and the other independent variables discussed here are defined and discussed more fully later in this chapter.

6. Peter M. Blau, *Exchange and Power in Social Life* (New York: John Wiley and Sons, 1964), p. 200.

7. Ibid., pp. 205–6.

8. Admittedly this is a personal impression. Nonetheless, I have spent over four years observing and hearing reports on Kenyan managerial behavior, especially in the civil service, and I am convinced that the statement is basically accurate.

9. The partial correlation coefficient between LAA Supervisory Effort and Visit Effort, controlling for the effect of the Size-Cohesion Interaction Effect, is $-.63$ ($d.f. = 11$; $p < .02$). We should note that it is very unlikely that the direction of causation is low Visit Effort leading to more Supervisory Effort. In that case, the supervisor would be observing the problem and then acting accordingly. The causal link would be direct and immediately obvious. In fact, the uncontrolled correlation between Visit Effort and LAA Supervisory Effort is quite weak ($r = -.25$; $N = 14$). We need to control for the Size-Cohesion Interaction Effect before the link becomes evident. It is unlikely that the decision by an LAA to increase his supervisory pressure would go through so complicated a preliminary filter. Other inverse correlations between closeness of supervision and productivity are reported in P. M. Blau and W. R. Scott, *Formal Organizations* (San Francisco: Chandler Publishing Company, 1962), p. 150.

10. Controlling for the Size-Cohesion Interaction Effect, the partial correlation coefficient between Visit Effort and priority to supervision is $-.54$, where that with frequency of supervision is $-.32$ ($d.f. = 11$; $p < .10$, $p < .30$ respectively).

11. Blau, *Exchange and Power in Social Life*, p. 143.

12. Ibid., p. 206.

13. Blau and Scott, *Formal Organizations*, pp. 124–25, 150; Amitai Etzioni, *Modern Organizations* (Englewood Cliffs, N.J.: Prentice-Hall, 1964), pp. 36–38; Sven Lundstedt, "Consequences of Reductionism in Organization Theory," *Public Administration Review* 32, no. 4 (July–August 1972): 328–33.

14. Leonard, ed., "Case Studies in Administrative Behaviour," chaps. 5, 20, 23. These studies cover a refectory, secondary schools, and coffee factories.

15. Sudhir Kakar, "Authority Patterns and Subordinate Behavior in Indian Organizations," *Administrative Science Quarterly* 16, no. 3 (September 1971): 298–307.

16. M. S. Makhanu and A. M. Ole, "Supervisory Styles in a Rural Soap Factory," in "Case Studies in Administrative Behaviour," ed. by Leonard, p. 22.8.

17. F. K. Muthaura, "Authoritarianism and Turnover in the East African Power and

Lighting Company," in "Case Studies in Administrative Behaviour," ed. by Leonard.

18. Lundstedt, "Consequences of Reductionism in Organization Theory."

19. Blau, *Exchange and Power in Social Life*, pp. 143–51, 205–6.

20. Jon R. Moris, "Managerial Structures and Plan Implementation in Colonial and Modern Agricultural Extension," in *Rural Administration in Kenya*, ed. D. K. Leonard (Nairobi: East African Literature Bureau, 1973), p. 113.

21. Elliott Jaques, *Measurement of Responsibility* (London: Tavistock Publications, 1956), esp. pp. 23, 32–42.

22. C. M. Ambutu et al., "The Effects of Supervisory Styles on Productivity in Cooperative Coffee Factories," in "Case Studies in Administrative Behaviour," ed. by Leonard.

23. R. H. O. Kowitti, "The Effects of Supervisory Style on Production in the Weaving Section of the Kisumu Cotton Mills," in "Case Studies in Administrative Behaviour," ed. by Leonard.

24. E. K. Arap-Bii et al., "Style and Effectiveness of Headmasters' Leadership in Secondary Schools," in "Case Studies in Administrative Behaviour," ed. by Leonard.

25. Ambutu et al., "The Effects of Supervisory Styles on Productivity in Cooperative Coffee Factories," p. 23.7.

26. Blau and Scott, *Formal Organizations*, p. 148.

27. Ambutu et al., "The Effects of Supervisory Styles on Productivity in Cooperative Coffee Factories," pp. 23.7–9.

28. Blau and Scott, *Formal Organizations*, pp. 155–56, 162–63. J. N. Wachira, "Hierarchical Independence and Subordinate Loyalty in a Rural Kenyan Bank," in "Case Studies in Administrative Behaviour," ed. by Leonard.

Chapter 6

1. See also E. R. Watts, "Measures to Increase Extension Effectiveness" (paper presented at the Universities of East Africa Social Sciences Conference, Dar es Salaam, December 27–31, 1970), p. 4.

2. The $r = .17$, $N = 169$, $p < .05$ in a two-tailed test. A correlation coefficient can be obtained for categorized data by assigning all those in the category (in this case, all those who had not fully completed primary school) a value of 1 on the independent variable and all others a value of 0. The advantage of this procedure is that it enables one to do multiple regression analysis and thus control for large numbers of independent variables. The alternate technique for this kind of data—analysis of variance—allows control for only one independent variable, at least on the computer facilities available to us in Nairobi.

3. The $r = .20$, $N = 169$, $p < .01$ in a two-tailed test for Agricultural Informedness. The $r = .15$, $N = 169$, $p < .05$ in a two-tailed test for the total of Explanatory Ability on all agricultural topics.

4. The correlations between various levels of education and Innovativeness are weak and not statistically significant. The peak performance level is reported here only because it conforms to the general tendency for more complex skills to demand higher levels of education.

5. R. K. Harrison, "Work and Motivation: A Study of Village-Level Agricultural Extension Workers in the Western State of Nigeria," mimeographed (Ibadan: Nigerian Institute of Social and Economic Research, 1969), pp. 228–29. Kidd, another student of

extension workers in Western Nigeria, also acknowledges this problem: David W. Kidd, "A Systems Approach to Analysis of the Agricultural Extension Service of Western Nigeria'" (Ph.D. dissertation, Department of Extension Education, University of Wisconsin, 1971), p. 64.

6. Harrison, "Work and Motivation," p. 258.

7. Ibid., pp. 72-83; Milton L. Blum and James C. Naylor, *Industrial Psychology: Its Theoretical and Social Foundations* (London: Harper and Row, 1968), pp. 328-63.

8. James G. March and Herbert A. Simon, *Organizations* (New York: John Wiley and Sons, 1958), p. 48.

9. Victor H. Vroom, *Work and Motivation* (New York: John Wiley and Sons, 1964), p. 186.

10. March and Simon, *Organizations*, p. 84.

11. This formulation refers to figure 3 and is a precise statement of the relationship described in the preceding sentence. It means that the second variable (4.1) has a negative effect on the first variable (4.2). The variables are those specified in the key to figure 3. If the hypothesized relationship had been positive rather than negative, a plus sign (+) would have been used instead of a minus.

12. Robert M. Price, "Organizational Commitment and Organizational Character in the Ghanaian Civil Service" (paper presented at the Social Science Research Council Conference on the Development of African Bureaucracies, Belmont, Maryland, March 6-9, 1974), p. 17.

13. Blum and Naylor, *Industrial Psychology*, pp. 342-43.

14. Abraham H. Maslow, *Motivation and Personality*, 2d ed. (New York: Harper and Row, 1970), pp. 35-47.

15. G. D. A. Owinoh, "The Effects of the Assembly Line Mode of Production on Worker Morale and Behaviour at the Limuru Bata Shoe Factory," in "Case Sudies in Administrative Behaviour," ed. by D. K. Leonard, mimeographed (Nairobi: Department of Government: University of Nairobi, 1973), pp. 16.8-9.

16. Douglas McGregor, *The Human Side of Enterprise* (New York: McGraw-Hill, 1960), p. 40.

17. Harrison, "Work and Motivation," p. 98.

18. Ibid., pp. 288-89; David W. Kidd, "Factors Affecting Farmers' Response to Extension in Western Nigeria," CSNRD-30, mimeographed (East Lansing, Mich.: Consortium for the Study of Nigerian Rural Development, 1968), p. 75.

19. F. K. Muthaura, "Authoritarianism and Turnover in the East African Power and Lighting Company," in "Case Studies in Administrative Behaviour," ed. Leonard.

20. Harrison, "Work and Motivation," pp. 152-62.

21. Rose Gakuya in David K. Leonard with G. S. Fields, R. W. Gakuya, and H. W'Opindi, "A Survey Evaluating the Curriculum of the Animal Health and Industry Training Institute," mimeographed (Nairobi: Institute for Development Studies, University of Nairobi, 1971), pp. 66-67.

22. Peter Marris and Anthony Somerset, *African Businessmen* (London: Routledge and Kegan Paul, 1971), p. 66.

23. The correlation of secondary education with job satisfaction is $r = -.14$ (N = 169, $p < .05$) and with career commitment is $r = -.23$ (N = 169, $p < .05$).

24. Amitai Etzioni, *Modern Organizations* (Englewood Cliffs, N.J.: Prentice-Hall, 1964), p. 40.

25. March and Simon, *Organizations*, p. 48; Vroom, *Work and Motivation*, p. 186.

26. These unrealistic expectations have been caused by first the acceleration and then the completion of Africanization. After Independence there was tremendous pressure to fill with Kenyan Africans posts that formerly had been held by Europeans and Asians. While the Africanization process was going on, virtually anyone with the minimum formal qualifications was able to get a good job. Once the process approached completion, however, only a few jobs were available to new entrants into the market, and even those with very high qualifications had difficulty finding a job. In the first stages of Africanization, all those with secondary education benefited. Now the process has been concluded at that level and is nearly complete for holders of general university degrees as well. Only those with professional or other highly specialized qualifications are still able to replace expatriates and thus remain in great demand.

27. Based on class discussions with students at the University of Nairobi.

28. Vroom, *Work and Motivation*, pp. 15–18.

29. Since a promotion system involves channeling personnel toward the top of the organization, it is critical that some positions there remain to be filled. There are a number of such positions still remaining in the technical ministries, but they will soon be filled if the present rapid expansion in university education continues. There are relationships between promotion structures and national manpower planning which lead us into difficult political issues. We will return to these in chapter 12.

30. Harrison, "Work and Motivation," pp. 155–61.

31. Leonard, "Organizational Structures for Productivity in Agricultural Extension," in *Rural Administration in Kenya*, ed. D. K. Leonard (Nairobi: East African Literature Bureau, 1973), pp. 151–52.

32. Informal conversations with AAs, AHAs, Ian Wallace (Principal of Embu), and Igor Mann (Project Director at AHITI) during 1971.

33. Blum and Naylor, *Industrial Psychology*, pp. 345–52.

34. Vroom, *Work and Motivation*, p. 18.

35. E. E. Lawler, "The Mythology of Management Compensation," mimeographed (New Haven: Yale University, 1965), cited in Blum and Naylor, *Industrial Psychology*, p. 353.

36. "Predominant" and "equal in importance" are determined as follows: for each individual, the factor he rates first in importance is given 4 points; the second, 3; the third, 2; and the fourth, 1. When 7 or more of the 10 possible points are held by a particular set of factors, we call it "predominant." When 4 to 6 are held by that set, we call it "equal in importance."

37. Jon R. Moris, "Managerial Structures and Plan Implementation in Colonial and Modern Agricultural Extension," in *Rural Administration in Kenya*, ed. by Leonard, p. 113.

38. Norman Uphoff, "African Bureaucracy and the 'Absorptive Capacity' for Foreign Development Aid: Analysis and Suggestions, With Considerations of the Case of Ghana" (paper presented at the Social Science Research Council Conference on the Development of African Bureaucracies, Belmont, Maryland, March 6–9, 1974), p. 37.

Chapter 7

1. E. R. Watts, "Agricultural Extension in Embu District of Kenya," *East African Journal of Rural Development* 2 (1969): 69–71; David K. Leonard with Bernard Chahilu

and Jack Tumwa, "Some Hypotheses Concerning the Impact of Kenya Government Agricultural Extension on Small Farmers," mimeographed (Nairobi: Institute for Development Studies, University of Nairobi, 1970), pp. 6–7.

2. P. M. Blau and W. R. Scott, *Formal Organizations* (San Francisco: Chandler Publishing Company, 1962), pp. 116–39; David K. Berlo, *The Process of Communication* (San Francisco: Rinehart Press, 1960); Erwin P. Bettinghaus, *Persuasive Communication* (New York: Holt, Rinehart and Winston, 1968). Bettinghaus discusses communication in hierarchies in chapter 10 without either mentioning any studies or presenting a theoretical model. Herbert A. Simon, Donald W. Smithburg, and Victor A. Thompson, *Public Administration* (New York: Alfred A. Knopf, 1950), pp. 218–59, 572–74; these three conclude, "The problem of administrative communication needs not only systematic statement but also systematic research" (p. 574).

3. Berlo, *The Process of Communication*, pp. 23–105.

4. The number of cases on which the following figures are based is sometimes quite small, and so the figures should be seen as examples and not as sound estimates.

5. For an elaboration of this point, see David K. Leonard, "Communications and Deconcentration," in *Development Administration: The Kenyan Experience*, ed. by G. Hyden, R. Jackson, and J. Okumu (Nairobi: Oxford University Press, 1970), pp. 91–111.

6. We are grateful to economists David Black and Rob Engle (who were then at the University of Nairobi), who helped us think through the methodology of making these measurements.

7. These and the other percentages reported here represent the additional proportion of the total possible amount of Agricultural Informedness or Explanatory Ability added by having this particular characteristic. They are calculated by dividing the highest possible score given by the full regression equation into the difference in scores given by having the highest and lowest amounts of the particular characteristic. Other methods of calculating and describing the magnitude of impact of these independent variables could be used, but all indicate approximately the same relative importance. (In fact another method was used in Leonard, "Organizational Structures for Productivity in Agricultural Extension," pp. 135 ff.) We offer this particular set of percentages because they seem to be conceptually the most direct and uncomplicated.

8. This calculation is inspired by a similar formula developed by Wilbur Schramm. Fraction of Selection = Expected Reward / Expected Energy Required. Wilbur Schramm, ed., *The Process and Effects of Mass Communication* (Urbana: University of Illinois Press, 1954), p. 19.

9. Carl I. Hovland, Irving L. Janis, and Harold H. Kelley, *Communication and Persuasion* (New Haven: Yale University Press, 1953), p. 246.

10. Ibid., p. 247.

11. This proposal is made more fully in Leonard, "Organizational Structures for Productivity in Agricultural Extension," pp. 140–41.

12. Hovland, Janis, and Kelley, *Communication and Persuasion*, pp. 36–39.

13. Ibid., p. 137.

14. Theodore Caplow, *Principles of Organization* (New York: Harcourt, Brace and World, 1964), p. 96.

15. David W. Kidd, "Factors Affecting Farmers' Response to Extension in Western Nigeria," CSNRD-30, mimeographed (East Lansing, Mich.: Consortium for the Study of Nigerian Rural Development, 1968), p. 91.

16. The single case of no apparent effect is spurious, caused by the inclusion of LAA's Informedness in the equation. Without it the item would add approximately 5 percent.

17. Bettinghaus, *Persuasive Communication*, pp. 200–202.

18. Ibid., pp. 211–12.

19. Kenya, Ministry of Agriculture, "Final Report of the Working Party on Agricultural Extension Services," mimeographed (Nairobi: 1970), p. 15. I was a member of this working party and my preliminary research findings contributed to the recognition of this need.

20. Niels Röling, Salome Kiarie, Gatenjwa Kuria, and Joshua Muiru, "Evaluation of JAA Training and Manual: Views of FTC Teachers and JAAs: Report to the Head of the Training Division Ministry of Agriculture" (Nairobi: Institute for Development Studies, University of Nairobi, 1972).

21. We are grateful to Niels Röling, formerly of the Institute for Development Studies, University of Nairobi, for these and other pieces of information on the current operation of the ministry's information system.

22. Donald Luebbe, Extension Adviser to the ministry from 1970 to 1972, took this point of view in discussions with us and in his recommendations.

23. Kenya, Ministry of Agriculture, "Final Report of the Working Party on Agricultural Extension Services," pp. 4–5.

Chapter 8

1. P. M. Blau and W. R. Scott, *Formal Organizations* (San Francisco: Chandler Publishing Company, 1962), pp. 116–28, esp. p. 125.

2. The LAAs report an average of 3.29 contacts with their subordinates per month. The subordinates report an average of 3.02.

3. The AAOs report an average of 2.01 contacts with their subordinates in a month while the junior staff report an average of 2.79.

4. W. Ouma Oyugi, "Participation in Development Planning at the Local Level," in *Rural Administration in Kenya*, ed. by D. K. Leonard (Nairobi: East African Literature Bureau, 1973), pp. 53–75.

5. Philip M. Mbithi, "Agricultural Extension as an Intervention Strategy," in *Rural Administration in Kenya*, ed. by Leonard, pp. 76–96.

6. Leonard Joy as quoted in Guy Hunter and Janice Jiggins, eds., *Stimulating Local Development* (London: Overseas Development Institute, 1976), p. 22.

7. This estimate was made with a small sample; at the 95 percent confidence level, it can be said that the true figure lies between none and 15 percent of the farmers. Even if the larger figure were accepted, the senior staff perception is false. David K. Leonard with Bernard Chahilu and Jack Tumwa, "Some Hypotheses Concerning the Impact of Kenya Government Agricultural Extension on Small Farmers," mimeographed (Nairobi: Institute for Development Studies, University of Nairobi, 1970), p. 6.

8. Jon R. Moris, "Multi-Subject Farm Surveys Reconsidered: Some Methodological Lessons" (paper presented at the East African Agricultural Economics Society Conference, Dar es Salaam, April 1–4, 1970), p. 9.

9. Blau and Scott, *Formal Organizations*, pp. 116–28.

10. Paul Devitt as quoted in Hunter and Jiggins, eds., *Stimulating Local Development*, p. 18.

11. Robert Chambers, *Managing Rural Development: Ideas and Experience from East Africa* (Uppsala: The Scandanavian Institute of African Studies, 1974), p. 133.

12. Robert Chambers, *Settlement Schemes in Tropical Africa* (London: Routledge and Kegan Paul, 1969), pp. 55–58.

13. Jon R. Moris, "Managerial Structures and Plan Implementation in Colonial and Modern Agricultural Extension," in *Rural Administration in Kenya*, ed. by Leonard, pp. 109–10.

14. Hunter and Jiggins, eds., *Stimulating Local Development*, p. 29. Italics removed.

Chapter 9

1. R. J. M. Swynnerton, *A Plan to Intensify the Development of African Agriculture in Kenya* (Nairobi: Government Printer, 1955), p. 10.

2. Francine R. Frankel, *India's Green Revolution: Economic Gains and Political Costs* (Princeton, N.J.: Princeton University Press, 1971). E. M. Rogers, J. R. Ascroft, and N. G. Röling, *Diffusion of Innovations in Brazil, Nigeria and India* (East Lansing: Department of Communication, Michigan State University, 1970).

3. Portions of the following analysis are drawn directly from David K. Leonard, "The Social Structure of the Agricultural Extension Services in the Western Province of Kenya," *The African Review* 2, no. 2 (1972): 323–43.

4. We are extremely grateful to the ministry for making these data available to us. The analysis and interpretation are entirely our own and the views expressed should not be interpreted as reflecting those of the Agricultural Statistics Section or of the Kenya government.

5. This number, 637, excludes interviews conducted on settlement schemes and in the Northern division of Busia district. Neither of these had been included in our initial study of extension workers, and they were excluded here to give us comparable information for the two sets of material.

6. I am grateful to my colleague W. Ouma Oyugi for this insight, gained during his research in South Nyanza.

7. This estimate has been made by using our staff survey data to estimate the total number of extension visits made in a year's time and then dividing that figure by the number of families in the province as indicated by: Kenya, Ministry of Finance and Planning, Statistics Division, *Kenya Population Census 1969*, vol. 3 (Nairobi: Government Printer, 1970). The following average visit figures are derived from this estimate, our staff survey data on the proportion of visits given to each category of farmers, and the Agricultural Statistics survey data on the proportion of each of these groups of farmers in the population.

8. Joseph Ascroft, "The Tetu Extension Pilot Project," in *Strategies for Improving Rural Welfare*, ed. by M. E. Kempe and L. D. Smith (Nairobi: Institute for Development Studies, University of Nairobi, 1971), pp. 65–69; Solomon N. Njooro, "Distribution of Veterinary Services in Mathira Division of Nyeri District: A Case Study" (B.A. dissertation, Department of Government, University of Nairobi, 1973). H. U. E. Thoden van Velzen, "Staff, Kulaks and Peasant," in *Socialism in Tanzania*, vol. 2, *Policies*, ed. by Lionel Cliffe and John S. Saul (Nairobi: East African Publishing House, 1973), pp. 153–79.

9. Rogers, Ascroft, and Röling, *Diffusion of Innovations in Brazil, Nigeria and India*, pp. 4.53–55.

10. R. G. Saylor, "An Opinion Survey of Bwana Shambas in Tanzania," mimeographed (Dar es Salaam: Economic Research Bureau, University of Dar es Salaam, 1970), pp. 12, 17.

11. E.g.F.E. Emery and O. A. Oeser, *Information, Decision and Action* (Melbourne: Melbourne University Press, 1958).

12. The greater part of my understanding of these two sets of justifications has come from frequent discussions and arguments with those who offer them. I am particularly grateful for the intellectual stimulation of the economists and communication specialists in the Institute for Development Studies of the University of Nairobi.

13. E.g. E. M. Rogers, *The Diffusion of Innovations* (New York: Free Press, 1962).

14. Private communication from Jeffrey Steeves, who has done research on the Kenya Tea Development Authority.

15. Rogers, Ascroft, and Röling, *Diffusion of Innovations in Brazil, Nigeria and India*, pp. 4.53–55.

16. B. P. Sinha and Prayag Mehta, "Farmers' Need for Achievement and Change-Proneness in Acquisition of Information from a Farm-Telecast," *Rural Sociology* 37, no. 3 (1972): 421, 423.

17. David K. Leonard with Bernard Chahilu and Jack Tumwa, "Some Hypotheses Concerning the Impact of Kenya Government Agricultural Extension on Small Farmers," mimeographed (Nairobi: Institute for Development Studies, University of Nairobi, 1970), pp. 6, 7, 10–12, 13.

18. We are indebted to former chief Matthew Mwenesi for this particular point and to him and former locational clerk Benjamin Kapitain for confirming our intuition on this general problem.

19. Uma K. Srivastava, Robert W. Crown, and Earl O. Heady, "Green Revolution and Farm Income Distribution," *Economic and Political Weekly* (Bombay) 6 (December 1971): A163–72.

20. Philip Mbithi's theoretical analysis of the problems of change in rural Kenya supports this argument. (See Philip M. Mbithi, "Agricultural Extension as an Intervention Strategy," in *Rural Administration in Kenya*, ed. D. K. Leonard (Nairobi: East African Literature Bureau, 1973.) Some of the extension experiments in the current Kenyan Special Rural Development Programme should cast empirical light on these two propositions.

21. A recent ILO team also argued for providing more agricultural extension to the bulk of Kenyan farmers. International Labour Office, *Employment, Incomes and Equality: A Strategy for Increasing Productive Employment in Kenya* (Geneva, 1972), p. 153.

22. E.g. Tom Mboya, "Sessional Paper No. 10: It Is African and It Is Socialism," *East African Journal* 6, no. 5 (May 1969): 15, 16. Also, Kenya, *Development Plan 1970–1974* (Nairobi: Government Printer, 1969), pp. 1–3. The plan proclaims an interest in equitable income distribution but limits the means of achieving it to taxation and rural development.

23. These estimates are quite rough. The total cattle population for the province was derived by multiplying the average number of cattle per farm (as indicated by the Agricultural Statistics survey) by the number of families in the province (indicated in the 1969 *Census*, vol. 3). The overall average number of cattle seen per year was estimated by dividing the total number of cattle by the total number of veterinary calls made in

one year (estimated from our staff survey data). The staff survey data also indicate the proportion of visits made to each type of farmer and the Agricultural Statistics survey shows the proportions of cattle in the province belonging to each of these types. These ratios then enabled us to calculate the average number of cattle seen per year for each type of farmer owner. Note that the ratios of relative advantage of each of the types of farmers is more reliable than are the estimates of the average number of visits each can expect to receive.

24. Njooro, "Distribution of Veterinary Services in Mathira Division of Nyeri District," p. 16.

25. Thoden van Velzen, "Staff, Kulaks and Peasant."

26. Ascroft, "The Tetu Extension Pilot Project," p. 56.

27. Private communication from Peter Moock, Institute for Development Studies evaluator of the Special Rural Development Programme in Vihiga.

28. P. M. Blau and W. R. Scott, *Formal Organizations* (San Francisco: Chandler Publishing Company, 1962), pp. 104–7.

29. Colin Leys, *Underdevelopment in Kenya: The Political Economy of Neo-Colonialism* (Berkeley, University of California Press, 1974), pp. 52–53.

30. Ibid., pp. 105–14.

31. Robert Chambers and David Feldman, "Report on Rural Development," mimeographed (Gaborones: Ministry of Finance and Development, Botswana, 1972).

32. Joseph Ascroft, Niels Röling, Joseph Kariuki, and Fred Chege, *Extension and the Forgotten Farmer: First Report of a Field Experiment* (Wageningen: Afdelingen voor Sociale Wetenschappen aan de Landbouwhogeschool, 1973).

Chapter 10

1. Jon R. Moris, "Managerial Structures and Plan Implementation in Colonial and Modern Agricultural Extension," in *Rural Administration in Kenya*, ed. D. K. Leonard (Nairobi: East African Literature Bureau, 1973). Hereafter cited as "Managerial Structures"; Jon R. Moris, "Administrative Penetration as a Tool for Development Analysis: A Structural Interpretation of Agricultural Administration in Kenya" (paper presented at the Conference on Comparative Administration in East Africa, Arusha, Tanzania, September 1971); hereafter cited as "Administrative Penetration").

2. Moris, "Administrative Penetration," p. 33.

3. Ibid., p. 34.

4. Moris, "Managerial Structures," p. 116.

5. Robert Chambers, *Settlement Schemes in Tropical Africa* (London: Routledge and Kegan Paul, 1969), pp. 71–75; J. Gus Liebenow, "Agriculture, Education, and Rural Transformation—with Particular Reference to East Africa" (Bloomington: Department of Political Science, Indiana University, 1969), pp. 13–14.

6. This inertia can be documented from 1931 to 1960. Jon R. Moris, "Crop Introduction Campaigns as a Test of Planning Capability in Extension Administration: The Central Kenyan Experience" (unpublished paper, Dar es Salaam, 1972), p. 33 (hereafter cited as "Crop Introduction").

7. Moris, "Managerial Structures," p. 115.

8. Nineteen men (56 percent) stated that they worked harder under the Europeans, opposed to 8 (23 percent) who said they worked less hard.

9. Eighteen men (53 percent) said that the Europeans knew less, against 8 (23 percent) who said they knew more.

10. R. Cohen, "Conflict and Change in a Northern Nigerian Emirate," in *Explorations in Social Change*, ed. by G. K. Zollschan and W. Hirsch (New York: Houghton Mifflin, 1964), pp. 495–521.

11. Kenya, Ministry of Agriculture, "Final Report of the Working Party on Agricultural Extension Services," mimeographed (Nairobi: 1970), p. 3.

12. Both approaches are evidenced in their classic form in a circular issued by District Agricultural Officer Nyeri on July 17, 1972, Ref: STAFF/4/25.

13. Jeffrey Steeves, private communication (Nairobi 1971). Steeves did extensive field research on the KTDA during 1970–1971.

14. Ibid.

15. Moris, "Crop Introduction," p. 32.

16. P. M. Blau and W. R. Scott, *Formal Organizations* (San Francisco: Chandler Publishing Company, 1962), p. 178.

17. For example, see Philip Mbithi, "Agricultural Extension as an Intervention Strategy," in *Rural Administration in Kenya*, ed. by Leonard, pp. 83 ff.; International Labour Office, *Employment, Incomes and Equality: A Strategy for Increasing Productive Employment in Kenya* (Geneva, 1972), p. 153; G. D. Hursh, N. S. Röling, and G. B. Kerr, "Communication in Eastern Nigeria: An Experiment in Introducing Change," mimeographed (East Lansing: Department of Communication, Michigan State University, 1968), p. 6.

18. Kurt Lewin, "Group Decision and Social Change," in *Readings in Social Psychology*, ed. by T. M. Newcomb and E. L. Hartley (New York: Henry Holt, 1947), pp. 330–44.

19. Erastus S. Mbugua, Siegfried Schönherr, and Peter Wyeth, "Agricultural Extension and Farmers Training" in "The Second Overall Evaluation of the Special Rural Development Programme," mimeographed (Nairobi: Institute for Development Studies, University of Nairobi, 1975), pp. 8.14–16.

20. Ibid., p. 8.16.

21. Norman T. Uphoff and Milton J. Esman, *Local Organization for Rural Development: Analysis of Asian Experience* (Ithaca, N.Y.: Center for International Studies, Cornell University, 1974), p. 68.

22. Mbugua, Schönherr, and Wyeth, "Agricultural Extension and Farmers Training," p. 8.13.

23. Ibid., pp. 8.13, 8.27.

24. Robert Chambers, *Managing Rural Development: Ideas and Experience from East Africa* (Uppsala: The Scandanavian Institute of African Studies, 1974), p. 66.

25. In Kenya, "villages" are only administrative units and not a concentration of houses. In the small-farm areas, settlement is usually spread fairly evenly over the whole area.

26. Deryke Belshaw and Robert Chambers experimented in Mbere Division, Embu District, with a set of procedures for planning the work of junior staff on an individual visit basis. The system involved no validation of agent reports and so was not an inspection system. Its main advantage was that it enabled the divisional AAO to plan his extension priorities clearly and to set farm visit targets in consultation with his staff. As a

planning system it appeared to be successful and greatly appreciated by the AAOs who used it. See Chambers, *Managing Rural Development*, chap. 3.

27. See Michel Crozier, *The Bureaucratic Phenomenon* (Chicago: The University of Chicago Press, 1964), pp. 40–46.

28. Blau and Scott, *Formal Organizations*, pp. 178–79.

29. Herbert Chabala, David Kiiru, and Soloman Mukuna, "A Further Evaluation of the Programming and Implementation Management (PIM) System," mimeographed (Nairobi: Institute for Development Studies, University of Nairobi, 1973), pp. 11–12.

30. Amitai Etzioni, *Modern Organizations* (Englewood Cliffs, N.J.: Prentice-Hall, 1964), p. 38.

31. Chabala, Kiiru, and Mukuna, "A Further Evaluation of the Programming and Implementation Management (PIM) System," pp. 14–15; E. K. Arap-Bii et al., "Style and Effectiveness of Headmasters' Leadership in Secondary Schools," in "Case Studies in Administrative Behaviour," ed. by D. K. Leonard, mimeographed (Nairobi: Department of Government, University of Nairobi, 1973), pp. 20.13–15.

32. Chambers, *Managing Rural Development*, p. 68.

33. Uphoff and Esman, *Local Organization for Rural Development*, pp. xi–xii.

34. The KTDA and the British-American Tobacco Company outgrowers scheme have been notable successes. The coffee authority did very well until world coffee prices collapsed. The cotton authority, on the other hand, has had consistent troubles since the Second World War because of weak prices.

35. Guy Hunter and Janice Jiggins, eds., *Stimulating Local Development* (London: Overseas Development Institute, 1976), p. 6.

36. Uphoff and Esman, *Local Organization for Rural Development*, p. xix.

37. These and other observations about Tanzania grow out of my current research on decentralization there.

38. Chambers, *Managing Rural Development*, pp. 35–54, 176–89.

39. Chabala, Kiiru, and Mukuna, "A Further Evaluation of the Programming and Implementation Management (PIM) System"; Martin David, Walter Oyugi, and Malcom Wallis, "SRDP as an Experiment in Development Administration" in "The Second Overall Evaluation of the Special Rural Development Programme," pp. 19.4–11.

40. The Kenya Tea Development Authority experience is described by Moris in his "Managerial Structures," pp. 121–24.

41. The argument for the primacy of procedures is made by Robert Chambers in his "Planning for Rural Areas in East Africa: Experience and Prescriptions" in *Rural Administration in Kenya*, ed. by Leonard. In fact, many of the book's articles take this point of view. The procedural experiments are described in Chambers, *Managing Rural Development*.

42. Kenya, *Report of the Commission of Inquiry (Public Service and Remuneration Commission) 1970–71*, D. N. Ndegwa, Chairman (Nairobi: Government Printer, 1971), pp. 85–92.

43. Tom Burns and G. M. Stalker, *The Management of Innovation* (London: Tavistock Publications, 1961), esp. pp. 119–22.

44. Victor A. Thompson, "Administrative Objectives for Development Administration," *Administrative Science Quarterly* 9, no. 1 (1969): 91–108.

45. E.g. B. B. Schaffer, "The Deadlock in Development Administration," in *Politics and Change in Developing Countries*, ed. by Colin Leys (London: Cambridge University Press, 1969), pp. 192–93.

46. See Goran Hyden, "Social Structure, Bureaucracy and Development Administration in Kenya," *The African Review* 1, no. 3 (1972): 126.

47. Herbert Kaufman, *The Forest Ranger: A Study in Administrative Behavior* (Baltimore: Johns Hopkins University Press, 1960), pp. 48–56.

48. Blau and Scott, *Formal Organizations*, pp. 178–79; Crozier, *The Bureaucratic Phenomenon*, pp. 107–108.

49. Robert K. Merton, *Social Theory and Social Structure*, rev. ed. (Glencoe, Ill.: Free Press, 1957), pp. 195–206.

50. Moris, "Managerial Structures," p. 113.

51. Hyden, "Social Structure, Bureaucracy and Development Administration in Kenya," pp. 127–28.

52. Chabala, Kiiru, and Mukuna, "A Further Evaluation of the Programming and Implementation Management System," pp. 12–13. Chambers, *Managing Rural Development*, chap. 2.

53. Kaufman, *The Forest Ranger*, pp. 234–38.

54. Moris, "Managerial Structures," p. 124.

55. A central management office to perform such a function has been officially suggested in Kenya. Kenya, *Report of the Commission of Inquiry*, D. N. Ndegwa, chairman, p. 140.

56. See Geoffrey S. Kuria, "The District Officer: A Colonial Shell," in *Rural Administration in Kenya*, ed. Leonard. Also see chapter 6 of this book.

Chapter 11

1. Here as elsewhere in this summary analysis, we are assuming that all other major causal factors are equal (or controlled).

2. Robert Chambers, "Planning for Rural Areas in East Africa," in *Rural Administration in Kenya*, ed. D. K. Leonard (Nairobi: East African Literature Bureau, 1973), pp. 24–25.

3. The term *utilitarian* is Amitai Etzioni's, who proposes that there are also coercive and normative types of organizations. Amitai Etzioni, *A Comparative Analysis of Complex Organizations* (New York: The Free Press, 1961), pp. 12, 23.

4. For an analysis of five alternative foci in the study of comparative administration, see Warren F. Ilchman, *Comparative Public Administration and "Conventional Wisdom,"* Sage Professional Papers in Comparative Politics, number 01–021 (Beverly Hills, Calif.: Sage Publications, 1971).

5. For these purposes we considered a correlation coefficient greater than .10 and in the expected direction as supporting the hypothesis. We took coefficients greater than .20 in the direction opposite from that expected as indicating that the relationship did not exist. We considered coefficients between these parameters as indeterminate. This does not mean that these findings are statistically significant. As we were dealing with grouped data and had only fourteen groups, we needed high correlation coefficients to

achieve statistical significance. In only seven cases were hypotheses accepted with statistical significance.

6. Stanley E. Seashore, *Group Cohesiveness in the Industrial Work Group* (Ann Arbor: Institute for Social Research, University of Michigan, 1954), pp. 63–80.

7. Theodore Caplow, *Principles of Organization* (New York: Harcourt, Brace and World, 1964), p. v.

8. In a working paper that we wrote after collecting but before analyzing our data, we reduced Caplow's verbal model to a list of hypotheses about the character of a set of regression equations. We then made numerous unsuccessful attempts to find a group of indices and a set of regression equations that would fit the model. The closest we came is the equation given in table 16. As can be seen, this set of relationships is far removed from Caplow's model and the discovery of the set was not assisted by his predictions.

9. Okot p'Bitek, the Ugandan anthropologist, is presently writing a book along these lines. See also, Goran Hyden, *Efficiency Versus Distribution in East African Cooperatives* (Nairobi: East African Literature Bureau, 1973), p. 205.

10. See, for example, Theodore Schultz, *Transforming Traditional Agriculture* (New Haven: Yale University Press, 1964); Michael Lipton, "The Theory of the Optimizing Peasant," *The Journal of Development Studies* 4, no. 3 (1968): 327–51; M. Schluter and John W. Mellor, "New Seed Varieties and the Small Farm," *Economic and Political Weekly* (Bombay) 7 (March 1972): A31–38; Uma K. Srivastava and Vishnuprasad Nagadevara, "On the Allocative Efficiency under Risk in Transforming Traditional Agriculture," *Economic and Political Weekly* (Bombay) 7 (June 1972): A73–78.

11. The terminology here is that suggested by Hans L. Zetterberg, *On Theory and Verification in Sociology*, 3d ed. (Totowa, N.J.: Bedminster Press, 1965), p. 71.

12. An ideal case for such research would be an organization in which significant numbers of both black and white Americans held both managerial and subordinate positions and in which there was a fair amount of racial conflict.

13. Throughout this analysis we will use the term *ethnic group* to apply to all types of groups based on the assumption of some common ancestry, however mythical it may be. In some settings the ethnic groups might be racial groups; in others, tribes; in still others, subtribes, clans, or lineages. All have essentially the same implications for our analysis.

14. Issac Gobanga, "Hierarchical Independence, Subordinate Loyalty, and Power in a Kenyan Rural Bank," in "Case Studies in Administrative Behaviour," ed. D. K. Leonard, mimeographed (Nairobi: Department of Government, University of Nairobi, 1973).

15. D. Muigai and S. Njooro, "Authority and Class Consciousness in the Nyeri Ministry of Works Mechanical Workshop," in "Case Studies in Administrative Behaviour," ed. Leonard.

16. J. N. Mukuha, "Power Relationships in a Nairobi Bank," in "Case Studies in Administrative Behaviour," ed. Leonard.

17. M. S. Ng'ang'a and B. M. F. Mwangola, "Tribe, Subordinate Loyalty, and Power in a Section of the Government Bureaucracy," in "Case Studies in Administrative Behaviour," ed. Leonard.

18. K. R. J. Sandbrook, "Politics in Emergent Trade Unions: Kenya 1952–1970" (Ph.D. dissertation, University of Sussex, 1970), pp. 498, 504.

19. For related evidence on cleavages in cooperatives, see John Saul, "Marketing Cooperatives in a Developing Country: The Tanzanian Case," in *University of East Africa Social Science Council 5th Annual Conference Proceedings* (Nairobi: University of Nairobi, 1970), pp. 586–601.

20. Max Weber, *From Max Weber: Essays in Sociology*, trans. and ed. by H. H. Gerth and C. W. Mills (New York: Oxford University Press, 1958), p. 203.

21. T. Burns and G. M. Stalker, *The Management of Innovation* (London: Tavistock Publications, 1961).

22. This view is shared by Neil J. Smelser, *Essays in Sociological Explanation* (Englewood Cliffs, N.J.: Prentice-Hall, 1968), pp. 74–75.

23. This is not to say that money presents no difficulties in the cross-cultural assessment of values, only that the considerable problems it poses are less than those raised by the units of measurement used by the other social sciences. See Smelser, *Essays in Sociological Explanation*, pp. 66–68.

24. James G. March and Herbert A. Simon, *Organizations* (New York: John Wiley and Sons, 1958), p. 84.

25. Victor H. Vroom, *Work and Motivation* (New York: John Wiley and Sons, 1964), pp. 15–18.

26. Peter M. Blau, *Exchange and Power in Social Life* (New York: John Wiley and Sons, 1964), p. 4.

27. March and Simon, *Organizations*, p. 84.

28. Blau, *Exchange and Power in Social Life*, p. 143.

29. Ng'ang'a and Mwangola, "Tribe, Subordinate Loyalty, and Power in a Section of the Government Bureaucracy."

30. Michel Crozier, *The Bureaucratic Phenomenon* (Chicago: The University of Chicago Press, 1964), p. 210. Crozier's contribution to transcultural organization theory is found in pages 144–208.

Chapter 12

1. Kenya had an average per capita annual income of U.S. $120 in 1967, and in 1969 it was anticipated that this would be U.S. $154 by 1974. Between 1964 and 1968 the economy grew at an average rate of 6.3 percent per year. Kenya, *Development Plan 1970–1974* (Nairobi: Government Printer, 1969), p. 1.

2. Cherry Gertzel, *The Politics of Independent Kenya* (Nairobi: East African Publishing House, 1970), pp. 32–72.

3. Kenya, *Report of the Commission of Inquiry (Public Service and Remuneration Commission) 1970–71*, D. N. Ndegwa, Chairman (Nairobi: Government Printer, 1971), pp. 1–5.

4. Kenya, *Development Plan 1970–1974*, p. 31.

5. Ibid.

6. Hans Ruthenburg, *African Agricultural Production Development Policy in Kenya 1952–1965* (Berlin: Springer-Verlag, 1966), p. 47.

7. Kenya, *Development Plan 1970–1974*, p. 31.

8. Jon R. Moris, "Managerial Structures and Plan Implementation in Colonial and Modern Agricultural Extension" in *Rural Administration in Kenya*, ed. D. K. Leonard (Nairobi: East African Literature Bureau, 1973), p. 116.

9. Hans Ruthenberg, ed., *Smallholder Agriculture and Development in Tanzania* (Berlin: Springer-Verlag, 1968), p. 349.

10. It is important to note that these figures are based on development and not on what is called recurrent expenditures. The latter include the important item of salaries of staff holding established positions. The recurrent expenditures are not itemized by purpose, so research expenditures cannot be factored out of them. There also have been some changes in how the *Estimates* have been presented over the years, and we have made some adjustments to make the figures comparable. The 1963–1964 development budget included contributions to the recurrent budget, and we have deducted these figures in making our calculations. In addition it should be noted that some aid contributions to research are not made through the budget and are not recorded in the *Estimates*. This tendency appears to have increased since Independence, so actual development expenditure on research in 1970–1971 may be somewhat better than we have indicated. The percentage differences are so great, however, that the general point we are making seems to be still valid. Kenya, *Development Estimates for the Year 1958-59* (Nairobi: Government Printer, 1958), pp. 7–9, 19–25; Kenya, *Development Estimates for the Year 1963-64* (Nairobi: Government Printer, 1963), pp. ii, 20–22; Kenya, *Development Estimates for the Year 1970-71* (Nairobi: Government Printer, 1970), pp. i, 1–3.

11. See Moris, "Managerial Structures," pp. 121–24; D. M. Etherington, "Economies of Scale and Technical Efficiency: A Case Study in Tea Production," *East African Journal of Rural Development* 4, no. 1 (1971): 72–87.

12. Philip Mbithi, "Agricultural Extension as an Intervention Strategy," in *Rural Administration in Kenya*, ed. Leonard, p. 81.

13. Geoffrey S. Kuria reaches a similar conclusion in his argument that the behavioral structure of Kenya's Provincial Administration causes it to waste its human resources. Kuria, "The District Officer: A Colonial Shell," in *Rural Administration in Kenya*, ed. Leonard, pp. 39–52.

14. See Y. P. Ghai and J. P. W. B. McAuslan, *Public Law and Political Change in Kenya* (Nairobi: Oxford University Press, 1970), pp. 79–124.

15. We use the term *European* to signify all whites present in Kenya whatever their national origins. This is consistent with Kenyan English usage.

16. Here and in the following discussions we use the term "Africans" as a shorthand for Kenyan Africans.

17. See Frank Holmquist, "Implementing Rural Development Projects," in *Development Administration: The Kenyan Experience*, ed. by G. Hyden, R. Jackson, and J. Okumu (Nairobi: Oxford University Press, 1970), pp. 201–29.

18. Goran Hyden, *Efficiency versus Distribution in East African Cooperatives* (Nairobi: East African Literature Bureau, 1973), pp. 142, 143, 217.

19. Kenya, *Development Plan 1970-1974*, p. 219.

20. Kenya, Ministry of Agriculture, "Final Report of the Working Party on Agricultural Extension Services," mimeographed (Nairobi: 1970).

21. Kenya, *Report of the Commission of Inquiry*, D. N. Ndegwa, chairman.

22. Kenneth Prewitt, "The Functional Justification of Inequality and the Ndegwa Report" (paper presented to the East African Social Science Council Conference, Nairobi, December 19–23, 1972).

23. The best examples are the experiences of Robert Chambers (*Managing Rural*

Development [Uppsala: The Scandanavian Institute of African Studies, 1974]), Joseph Ascroft ("The Tetu Extension Pilot Project," in *Strategies for Improving Rural Welfare*, ed. M. E. Kempe and L. D. Smith [Nairobi: Institute for Developmental Studies, University of Nairobi, 1971]), and Erastus S. Mbugua, Siegfried Schönherr, and Peter Wyeth ("Agricultural Extension and Farmers Training," in "The Second Overall Evaluation of the Special Rural Development Programme," mimeographed [Nairobi: Institute for Development Studies, University of Nairobi, 1975]).

24. International Labour Office, *Employment, Incomes and Equality in Kenya* (Geneva, 1972), pp. 236-41.

25. For a discussion of similar forces at work in India, see Susanne H. Rudolph and Lloyd I. Rudolph, eds., *Education and Politics in India: Studies in Organization, Society, and Policy* (Cambridge, Mass.: Harvard University Press, 1972), pp. 25-80.

26. E.g. Ann Seidman, "The Inherited Dual Economies of East Africa," in *Socialism in Tanzania*, vol. 1: *Politics*, ed. by L. Cliffe and J. S. Saul (Dar es Salaam: East African Publishing House, 1972), pp. 41-64.

27. See, "The Arusha Declaration: Socialism and Self-Reliance," in Julius K. Nyerere, *Freedom and Socialism / Uhuru na Ujamaa* (Dar es Salaam: Oxford University Press, 1968), pp. 231-50.

28. Fred W. Riggs, *Administration in Developing Countries: The Theory of Prismatic Society* (Boston: Houghton Mifflin, 1964), pp. 260-67; Fred W. Riggs, "Bureaucrats and Political Development: A Paradoxical View," in *Bureaucracy and Political Development*, ed. by Joseph LaPalombara (Princeton, N.J.: Princeton University Press, 1963), pp. 120-67.

29. The following argument parallels that of Hyden, "Social Structure, Bureaucracy, and Development Administration in Kenya," *The African Review* 1, no. 3 (January 1972): 118-29.

30. Gertzel, *The Politics of Independent Kenya*, pp. 166-73.

31. Hyden, *Efficiency versus Distribution in East African Cooperatives*, p. 216.

32. The term *representative bureaucracy* in the latter sense figures prominently in W. L. Warner, et al., *The American Federal Executive* (New Haven: Yale University Press, 1963).

Index

ACY-3274

8/95
gift

S
544.5
K4
L45
1977